Unit 3

Inspire Science

Earth and Space

Impacts on the Environment

Phenomenon: Why are these corals white?

This image depicts a portion of a coral reef where coral bleaching has occurred. Coral bleaching occurs when environmental conditions change, such as an increase in water temperature. This causes the corals to expel the algae that live in their tissues, which turns the corals white.

The Great Barrier Reef is the world's biggest single structure made by living organisms!

FRONT COVER: Ethan Daniels/WaterFrame/Getty Images. **BACK COVER:** Ethan Daniels/WaterFrame/Getty Images.

mheducation.com/prek-12

Send all inquiries to:
McGraw-Hill Education
STEM Learning Solutions Center
8787 Orion Place
Columbus, OH 43240

ISBN: 978-0-07-688284-7
MHID: 0-07-688284-5

Printed in the United States of America.

7 8 9 10 11 12 LWI 24 23 22 21

McGraw-Hill is committed to providing instructional materials in Science, Technology, Engineering, and Mathematics (STEM) that give all students a solid foundation, one that prepares them for college and careers in the 21st century.

Welcome to

Inspire Science

Earth and Space

Explore Our Phenomenal World

Learning begins with curiosity. Inspire Science is designed to spark your interest and empower you to ask more questions, think more critically, and generate innovative ideas.

Start exploring now!

Inspire Curiosity • Inspire Investigation • Inspire Innovation

Authors, Contributors, and Partners

Program Authors

Alton L. Biggs
Biggs Educational Consulting
Commerce, TX

Ralph M. Feather, Jr., PhD
Professor of Educational Studies and Secondary Education
Bloomsburg University
Bloomsburg, PA

Douglas Fisher, PhD
Professor of Teacher Education
San Diego State University
San Diego, CA

Page Keeley, MEd
Author, Consultant, Inventor of Page Keeley Science Probes
Maine Mathematics and Science Alliance
Augusta, ME

Michael Manga, PhD
Professor
University of California, Berkeley
Berkeley, CA

Edward P. Ortleb
Science/Safety Consultant
St. Louis, MO

Dinah Zike, MEd
Author, Consultant, Inventor of Foldables®
Dinah Zike Academy, Dinah-Might Adventures, LP
San Antonio, TX

Advisors

Phil Lafontaine
NGSS Education Consultant
Folsom, CA

Donna Markey
NBCT, Vista Unified School District
Vista, CA

Julie Olson
NGSS Consultant
Mitchell Senior High/Second Chance High School
Mitchell, SD

Content Consultants

Chris Anderson
STEM Coach and Engineering Consultant
Cinnaminson, NJ

Emily Miller
EL Consultant
Madison, WI

Key Partners

American Museum of Natural History
The American Museum of Natural History is one of the world's preeminent scientific and cultural institutions. Founded in 1869, the Museum has advanced its global mission to discover, interpret, and disseminate information about human cultures, the natural world, and the universe through a wide-ranging program of scientific research, education, and exhibition.

PhET Interactive Simulations
The PhET Interactive Simulations project at the University of Colorado Boulder provides teachers and students with interactive science and math simulations. Based on extensive education research, PhET simulations engage students through an intuitive, game-like environment where students learn through exploration and discovery.

SpongeLab Interactives
SpongeLab Interactives is a learning technology company that inspires learning and engagement by creating gamified environments that encourage students to interact with digital learning experiences. Students participate in inquiry activities and problem-solving to explore a variety of topics through the use of games, interactives, and video while teachers take advantage of formative, summative, or performance-based assessment information that is gathered through the learning management system.

Measured Progress, a not-for-profit organization, is a pioneer in authentic, standards-based assessments. Included with New York Inspire Science is **Measured Progress STEM Gauge®** assessment content which enables teacher to monitor progress toward learning NGSS.

Table of Contents

Impacts on the Environment

Module 1 Human Impact on the Environment

Module 2 Earth and Human Activity

Impacts on the Environment

Human Impact on the Environment

ENCOUNTER
THE PHENOMENON

How do humans impact Earth?

Urban Explosion

GO ONLINE
Watch the video *Urban Explosion* to see this phenomenon in action.

Communicate The top satellite image here shows Shanghai, China, in 1984. How does it compare to the bottom image from 2016? Record your ideas for how you think human activities impact Earth's land, water, atmosphere, and climate. Discuss your ideas with three different partners. Revise or update your ideas, if necessary, after the discussions with your classmates.

PERFORMANCE EXPECTATION

STEM Module Project Launch
Engineering Challenge

Lesson 1
Impact on Land

Lesson 2
Impact on Water

Lesson 3
Impact on the Atmosphere

Lesson 4
Impact on Climate

Who's moving in next door?

A developer would like to purchase a large area of farmland near a small town. She wants to build a shopping mall on the land. Community members are worried about the environmental impacts that the development may have on their area. They want to know how it will affect land, water, and air resources. A committee has been formed to research these impacts before any decisions are made.

You work for an environmental consulting firm. Your team's task is to design plans for monitoring and minimizing environmental impacts and submit those plans to the town committee.

Start Thinking About It

What do you know about the engineering design process? Discuss your thoughts with the class.

STEM Module Project

Planning and Completing the Engineering Challenge How will you meet this goal? The concepts you will learn throughout this module will help you plan and complete the Engineering Challenge. Just follow the prompts at the end of each lesson!

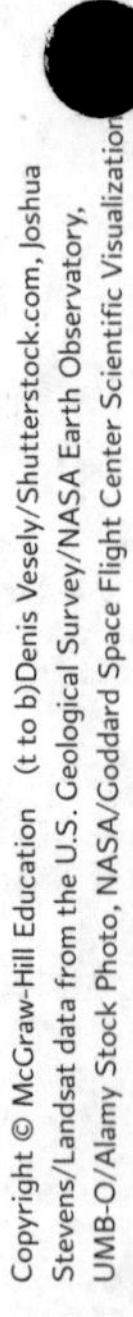

PAGE KEELEY SCIENCE PROBES

LESSON 1 LAUNCH

Environmental Pollution

Four friends were arguing about the source of pollution that harms the environment. They each had different ideas about where the pollution comes from that harms the environment. This is what they said:

Abe: I think harm to the environment comes from human activities.

Carrie: I think harm to the environment comes from natural processes.

Okee: I think harm to the environment comes from both human activities and natural processes.

Florio: I don't think it matters if harm to the environment comes from humans or natural processes. It depends on whether the source is non-biodegradable.

Circle the name of the friend you think has the best idea about sources of harm to the environment. Explain why you think that is the best idea.

You will revisit your response to the Science Probe at the end of the lesson.

LESSON 1

Impact on Land

ENCOUNTER THE PHENOMENON | How can fungi help minimize human impact on the land?

GO ONLINE
Watch the video *Mushroom Packaging* to see this phenomenon in action.

Record your observations about the phenomenon.

Now, brainstorm a list of five activities that you think might impact land on Earth. How might these activities also impact the biosphere? Next to each activity, list the potential negative effects as well as a potential counteractivity or technology that might monitor or minimize the impact.

EXPLAIN
THE PHENOMENON

Are you starting to get some ideas about how human activities can alter Earth's land resources and how we can mitigate these effects? Use your observations about the phenomenon to make a claim about the ways in which humans can negatively—and positively—impact Earth's land.

CLAIM

Solutions, like mushroom packaging, are important for Earth's land because...

COLLECT EVIDENCE as you work through the lesson.

Then return to these pages to record your evidence.

EVIDENCE

A. What evidence have you discovered to explain the impacts on land usage?

MORE EVIDENCE

B. What evidence have you discovered to explain how humans can monitor and minimize these impacts?

When you are finished with the lesson, review your evidence. If necessary, based on the evidence, revise your claim.

REVISED CLAIM

Solutions, like mushroom packaging, are important for Earth's land because...

Finally, explain your reasoning for how and why your evidence supports your claim.

REASONING

The evidence I collected supports my claim because...

How does a growing population impact Earth?

Scientists estimate that there were about 300 million humans on Earth a thousand years ago. Today there are more than 7 billion humans on Earth, as shown below. By 2050, there could be over 9 billion.

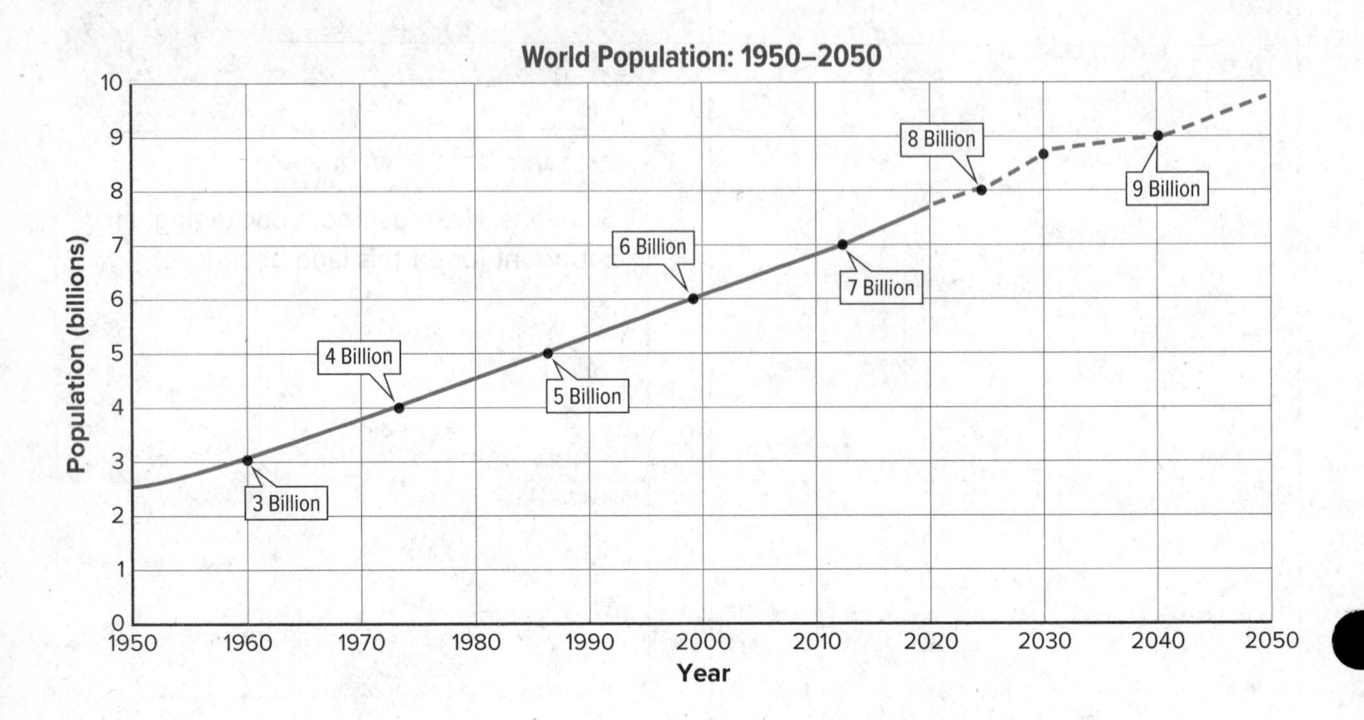

Typically as the human population increases, so does the consumption of natural resources. A **natural resource** is something from Earth that living things use to meet their needs. Every human being needs certain things, such as food, clean water, and shelter, to survive. As the human population grows, people need to build more houses and roads and clear more land for crops. Land itself is a resource. How does a growing population impact land as a resource?

Want more information?

Go online to read more about how the human population impacts land resources.

FOLDABLES

Go to the Foldables® library to make a Foldable® that will help you take notes while reading this lesson.

LAB Move Over

You're the first one on the school bus in the morning. After a few more stops, you notice that the bus is rather noisy. By the time you get to school, every seat is taken. Like the school bus, space on Earth is limited.

Materials

metric ruler

colored pencils

Procedure

1. In the Data and Observations section below, draw a square that is 10 cm on each side. This square represents 1 km^2 of land.
2. In 1965, an average of 22 people lived on 1 km^2 of land. Choose a color and draw 22 small circles inside your square to represent this.
3. In 1990, the average was 35 people per 1 km^2 of land. Choose a new color and add 13 circles to illustrate this increase.
4. In 2025, the estimated number of people will be 52 per 1 km^2 of land. Choose a new color and add enough circles to represent this increase.

Data and Observations

Procedure, continued

5. Prepare a bar graph that shows population density for these years. Plot population on the vertical axis and year on the horizontal axis.

Analyze and Conclude

6. Use your bar graph to explain changes in Earth's human population over time.

7. What impact might this have on land?

Land as a Resource No matter where you live, you and all living things use land for living space. Living space includes natural habitats, as well as the land on which buildings, sidewalks, parking lots, and streets are built.

People also use soil for growing crops and forests to harvest wood for making furniture, houses, and paper products. They mine minerals from the land, sometimes removing entire mountaintops, as shown to the right. In each of these cases, people use land to meet their needs.

What are the consequences of using land as a resource?

Humans sometimes cut forests to clear land for grazing, farming, or building houses or highways. **Deforestation** is the cutting of large areas of forests for human activities. What's the impact of deforestation?

INVESTIGATION

The Last Frontier Forests

1. **MATH Connection** The data in the table describe the difference between the size of Earth's total original forest, and the total remaining frontier forest. Calculate the percentage of the original forest that remains and enter that data into the fourth column of the table.

The World's Forests			
Geographical Area	Total Original Forest (in km^2)	Total Remaining Forest (in km^2)	Approximate % of Original Forest Remaining
Oceania	1,431,000	319,000	
Central America	1,779,000	172,000	
Africa	6,799,000	527,000	
South America	9,736,000	4,439,000	
North America	10,877,000	3,737,000	
Asia	15,132,000	844,000	
Russia/Europe	16,449,000	3,463,000	

ENVIRONMENTAL Connection Next, research how cutting old growth forests of North America's Pacific Northwest impacts the northern spotted owl and ultimately the biodiversity and viability of this natural system. Record your findings below.

Impact of Deforestation As you calculated in the previous activity, a significant amount of forests have been reduced globally. An increased need for resources produced by trees, or the land on which the trees grow, has led to a decrease in the amount of forests.

What's the impact? Deforestation leads to loss of animal habitats, which can lead to the endangerment or extinction of a species. In tropical rain forests—complex ecosystems that can take hundreds of years to replace—deforestation is a serious problem. Tropical rain forests are home to an estimated 50 percent of all species on Earth.

In addition, deforestation affects the atmosphere. Trees remove carbon dioxide from the atmosphere during photosynthesis. Rates of photosynthesis decrease when large areas of trees are cut down, and more carbon dioxide remains in the atmosphere.

People also clear land for development and agriculture. Let's investigate the impact of agriculture on land resources.

LAB Modeling Earth's Farmland

Safety

Materials

apple

plastic knife

Procedure

1. Read and complete a lab safety form.
2. Cut an apple into four quarters and set aside three. One quarter of Earth's surface is land. The remaining three-quarters of Earth's surface is covered with water.
3. Slice the remaining quarter into thirds.
4. Set aside two of the three pieces, because two-thirds of Earth's land is too hot, too cold, or too mountainous to farm or live on.
5. Carefully peel the remaining piece. This peel represents the usable land surface that must support the entire human population.
6. Follow your teacher's instructions for proper cleanup.

Analyze and Conclude

7. What might happen if available farmland is used for other purposes?

8. Why will the amount of farmland decrease in your lifetime?

9. How do you think we can grow more food on less land?

Land Usage for Agriculture As the human population grows, so does our need for food. Yet there are consequences for natural systems. Factors that affect agriculture needs such as pesticide usage, soil quality, and space can be detrimental to the production of food needed to support the world population. Examine the table below to learn more about how agriculture impacts the land.

Impacts of Agriculture	
Pesticide Usage To feed the growing world population, some farmers use higher-yielding seeds and chemical fertilizers. These methods help increase the amount of food grown on each km^2 of land. Herbicides and pesticides also are used to reduce weeds, insects, and other pests that can damage crops. However, runoff containing these fertilizers can seep into groundwater supplies, polluting drinking water. They can also run off into streams and rivers, affecting aquatic organisms.	
Soil Quality Whenever vegetation is removed from an area, such as tilled farmland, soil is exposed. Without plant roots to hold soil in place, nothing prevents the soil from being carried away by running water and wind. High rates of soil erosion can lead to desertification. Desertification is the development of desertlike conditions due to human activities and/or climate change. A region of land that undergoes desertification is no longer useful for food production.	
Space Today, about one-fifth of U.S. land is used for growing crops and about one-fourth is used for grazing livestock. Animals such as cattle eat vegetation and then are used as food for humans. About sixty-five percent of the land in Texas is used for grazing cattle. Other regions of the United States such as the west and Midwest also set aside land as pasture. Other land is used to grow crops to be fed to cattle. Many farmers raise corn and hay for this purpose.	

LIFE SCIENCE Connection Changes in biodiversity affect more than just one organism. Everything is connected, and, when one organism changes or disappears, then multiple organisms change or disappear. When an organism, such as a bee, faces changes in the environment, plants are affected because they are no longer being pollinated by the bees. This results in the production of crops with lower yields. The fewer crops there are being produced, the less food there is to feed the world's population.

Urbanization Large tracts of rural land in the United States have been developed as suburbs, or residential areas on the outside edges of a city and farther out into the country. The development of land for houses and other buildings near a city is called **urbanization.** What impact does development have on land resources? Let's investigate!

INVESTIGATION

Rainfall Runoff

It's not unusual for streams and rivers to flood after heavy rain. The amount of water flowing quickly into waterways may be more than streams and rivers can carry. Can changing the landscape increase flooding?

1. Observe as your teacher pours a bucket of water onto a paved area and one bucket of water onto grass. Discuss how the properties of the pavement and the grass affected the flow of the water. Record key points from your discussion below.

2. The table to the right lists the percentage of rainfall that runs off land. Compare the amount of runoff for each of the land uses listed. Assume that all of the regions are the same size and have the same slope. Looking at the table, do you see a relationship between what is on the land and how much water runs off of it?

Rainfall Runoff Percentages

Land Use	Runoff to Streams (%)
Commercial (offices and stores)	75
Residential (houses)	40
Natural Areas (forest and grassland)	29

3. Two years after construction of a mall near a stream, houses downstream flooded after a heavy rain. What do you think contributed to the flooding?

4. What are some ways that developers can reduce the risk of flooding?

Impact of Urbanization There are multiple environmental impacts related to urban development. As you just explored, paving land prevents water from soaking into the soil. Instead, it runs off into sewers or streams. A stream's discharge increases when more water enters its channel. Stream discharge is the volume of water flowing past a point per unit of time. During heavy rainstorms in paved areas, rainwater flows directly into streams, increasing stream discharge and the risk of flooding.

Urbanization can also cause habitat destruction, the loss of farmland, and negatively impact groundwater supplies. Many communities use underground water supplies for drinking. Covering land with roads, sidewalks, and parking lots reduces the amount of rainwater that soaks into the ground to refill underground water supplies.

Finally, many wetlands throughout the world have been drained and filled with soil for roads, buildings, airports, and housing developments. The disappearance of wetlands has also been associated with rising sea level, coastal erosion, and the introduction of species that are not naturally found in wetlands.

STEM Careers

A Day in the Life of a Wetland Specialist

Yuck! That's what many people think of wetlands. They think wetlands are murky swamps, filled with algae and teeming with snakes, alligators, and mosquitoes. They do not realize that wetlands are an extremely valuable resource and need to be protected from destruction. Here are just a few of their benefits:

- They provide habitat for thousands of species, including fish, mollusks, birds, reptiles, amphibians, and insects.
- Wetlands filter out natural and manufactured pollutants before they can enter rivers, lakes, or groundwater.
- Wetlands reduce storm and flood damage, prevent erosion, and restore groundwater and surface water.

What kind of scientist helps protect wetland areas? A wetland specialist! A wetland specialist assesses the health of a wetland area. He or she determines through field research how humans and natural activities influence wetland environments. The goal is to determine if significant changes have altered the area by tracking and monitoring the health of local vegetation, wildlife, and water levels. Extensive data of these areas are used to draft reports showing the health of a wetland system.

If a wetland area is damaged or contaminated, wetland specialists work to fix and restore the area to a healthy state. They remove pollutants from the wetland and implement remediation projects. They also track the introduction of invasive species that may be having a negative effect on the ecological balance of the wetland.

It's Your Turn

ENVIRONMENTAL Connection Research the wetland restoration in Point Reyes, California. Use the data you collected to design a poster to promote wetland conservation.

Read a Scientific Text

Urbanization has been a continuing trend for most of recorded history. In 2007, a threshold was reached in that more people lived in urban areas than in rural areas. What are the effects of urbanization on the biosphere?

CLOSE READING

Inspect

Read the passage *Urbanization Can Change the Relative Effects of Top-Down and Bottom-Up Factors.*

Find Evidence

Reread the article. Highlight the text explaining bottom-up and top-down systems.

Make Connections

Collaborate With your partner, cite evidence from the text that supports how changes to Earth's environments can have different impacts (both negative and positive) for different organisms. Why is understanding these impacts important?

Urbanization Can Change the Relative Effects of Top-Down and Bottom-Up Factors

Urbanization fragments habitat, introduces exotic species, and has effects far beyond the city boundaries. Stan Faeth and his colleagues (2005) compared species interactions in Phoenix, Arizona, and in the nearby rural Sonoran Desert area. In the Sonoran Desert, plant biomass is limited primarily by water availability. Associated herbivores are, in turn, limited mainly by plant quality and quantity. Natural enemies, such as avian predators, have relatively little effect on herbivore densities. This is a bottom-up driven system.

In urban Phoenix, plant biomass is much increased by watering. Even native Sonoran plants are generally more productive due to increased water usage, especially in dry periods when desert plants typically senesce. A higher and more consistently available plant biomass results from this increase in water resources. In turn, herbivores and their avian predators are more abundant. The increase in bird densities in suburban Phoenix translates into stronger top-down effects on herbivores. A long-term experiment in both Phoenix and the Sonoran Desert excluded birds from 40 brittlebush plants, *Encelia farinosa,* via the use of netting. The experiments were repeated in deserts, in desert remnants within the city, and in suburban yards. In the Sonoran Desert, bird exclusion had no effect on herbivore densities, which are limited by plant availability and water supply. Within the city, insect herbivores are more abundant and bird exclusion allowed herbivores to increase even more. With increased productivity, top-down factors had increased in importance. These results show how global change via urbanization can affect species interactions. These interactions are important because many direct and indirect services are provided by living organisms in urban environments, such as pollination of home garden plants by insects, control of pests by predators and parasites, and the recycling of nutrients.

ENVIRONMENTAL Connection How can urbanization influence the composition of natural systems?

Landfills and Hazardous Waste Land is also used when consumed products are thrown away. About 60 percent of our garbage goes into landfills. Some of these wastes are dangerous. Examine the table below to learn more about the impacts of landfills and hazardous waste.

Landfills	Hazardous Waste

About 34 percent of our trash is recycled and composted. About 11 percent is burned, and the remaining 55 percent is placed in landfills. Landfills are areas where trash is buried. Since many materials do not decompose in landfills, or they decompose slowly, landfills fill with garbage, and new ones must be built. Locating an acceptable area to build a landfill can be difficult. Type of soil, the depth to groundwater, and neighborhood concerns must be considered.	Some trash cannot be placed in landfills because it contains harmful substances that can affect soil, air, and water quality. This trash is called hazardous waste. The substances in hazardous waste also can affect the health of humans and other living things. Both industries and households generate hazardous waste. For example, hazardous waste from the medical industry includes used needles and bandages. Household hazardous waste includes used motor oil and batteries.

Pollution Runoff that contains chemicals from landfills, mineral mines, and agricultural fields can pollute and affect the quality of soil and water. **Pollution** is the contamination of the environment with substances that are harmful to life. Pollution can be devastating to many plant and animal species.

THREE-DIMENSIONAL THINKING

Explain what **effect** trash disposal can have on Earth's systems.

THREE-DIMENSIONAL THINKING

Summarize your understanding of the **cause-and-effect** relationships between human activities and the environmental impacts on land in the table below.

Type	Causes	Effects
Deforestation		
Agriculture		
Urbanization		
Waste Disposal		

COLLECT EVIDENCE

What are the impacts of land usage? Record your evidence (A) in the chart at the beginning of the lesson.

What actions help protect the land?

Earth's land is used in so many different ways. You have just learned about the impacts of each kind of land use, including pollution and biodiversity loss. Let's investigate what can be done to conserve land resources!

INVESTIGATION

Relaxing on the Refuse

Plastic refuse usually doesn't make an attractive seat. However, with a little effort, it can be reused in a variety of very useful ways. The bench you see here is made of reused plastic.

1. What kinds of items might be reused?

2. Describe how you reuse or recycle materials at home.

3. How might reusing a material help preserve the environment?

Resource Use Resources such as wood, petroleum, and metals are important for making the products you use every day at home and in school. However, if these resources are not used carefully, the environment can be damaged. **Conservation** is the careful use of Earth's materials to reduce damage to the environment. Conservation can prevent future shortages of some materials. How can individuals help manage land resources wisely?

INVESTIGATION

Trash Classification

1. Create a table with the columns *Paper, Plastic, Glass, Metal,* and *Food Waste* in the space below. Add a row for *Number of items* and one for *Rank.*

2. Record the items you throw out in one day. At the end of the day, count the number of trash items in each column.

3. Rank each column by the number of trash items from fewest to most.

4. How did your rankings compare with those of your classmates?

5. List two simple ways that you and your classmates can reduce your consumption of Earth's materials.

Reduce, Reuse, Recycle Developed countries such as the United States use more natural resources than other regions. Ways to conserve resources include reducing the use of materials, and reusing and recycling materials.

Reusing an item means finding another use for it instead of throwing it away. Using material again is called recycling. When you recycle wastes such as glass, paper, plastic, steel, or tires, you help conserve Earth's land resources. You can use yard waste and vegetable scraps to make rich compost for gardening, reducing the need for synthetic fertilizers. Compost is a mix of decayed organic material, bacteria, other organisms, and small amounts of water. Reducing means limiting the amount used initially.

The human population explosion already has had an effect on the environment and the organisms that inhabit Earth. It's unlikely that the population will begin to decline in the near future. To make up for this, resources must be used wisely. Conserving resources by reducing, reusing, and recycling is an important way that you can make a difference.

Managing Land Resources Because some land uses involve renewable resources while others do not, managing land resources is complex. In addition, the amount of land is limited, so there is competition for space. Landfills, for example, take up valuable space and often risk polluting the area. Those who manage land resources must balance all of these issues. Let's investigate how land resources can be managed and protected.

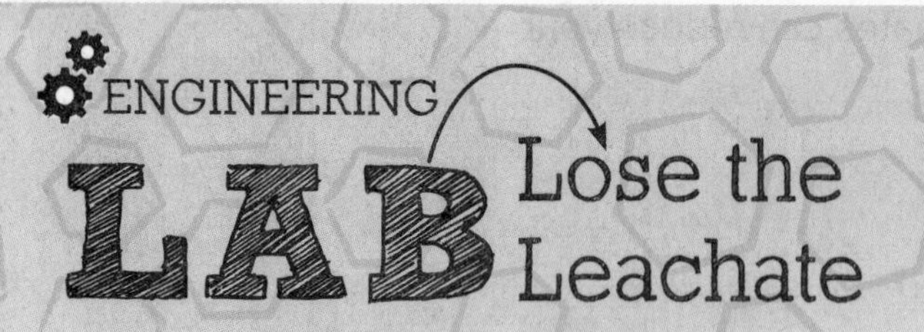

Safety 

Materials

creative construction materials

paper towels

scissors

masking tape

Procedure

1. Read and complete a lab safety form.
2. Read the information provided by your teacher about landfill requirements as set by the Environmental Protection Agency (EPA).
3. Plan and diagram your landfill design in your Science Notebook. Identify any materials used and note the inputs and outputs of the designed system.
4. Use the materials to build your landfill model. Add waste materials.
5. Pour 350 mL of water on your landfill to simulate rain. Observe the path the water takes.
6. Collect the leachate produced by your landfill. Leachate is the liquid that seeps out of your landfill. Record your data below.

7. Follow your teacher's instructions for proper cleanup.

Analyze and Conclude

8. Explain how you designed your model to function efficiently and meet requirements set by the EPA.

9. Identify and describe how each component of the proposed solution is intended to help with the amount of leachate produced by the landfill.

10. Compare your landfill design to at least three other groups. How did the amount of leachate produced by your landfill compare? Create a graph to organize your data compared to the other groups.

11. Analyze the data above. Identify and describe relationships among the datasets, including relationships between each team's design and how their design addressed the guidelines set forth by the EPA. What aspects of the design do you think best decreased the amount of leachate leakage? Write your response in your Science Notebook.

Analyze and Conclude, continued

12. Determine the similarities and differences among your design and the other groups'. In your evaluations, be sure to use data and/or digital tools and arguments to compare proposed solutions to your landfill. Identify the best characteristics of each that can be combined into a new solution to better address the effectiveness of the landfill. Redesign your landfill. Plan and diagram your new landfill in your Science Notebook. Retest your design.

13. How does your new designed solution better meet the needs of the landfill? Identify any limitations with your model in regards to representing the proposed solution.

14. Describe why it is important to test and modify solutions to designed systems, such as landfills.

Managing Land Resources One way governments can manage forests and other unique ecosystems is by preserving them. On preserved land, logging and development is either banned or strictly controlled. Large areas of forests cannot be cut. Instead, loggers cut selected trees and then plant new trees to replace the ones they cut. **Reforestation** involves planting trees to replace ones that have been removed. Trees are renewable—they can be replanted and grown in a relatively short amount of time.

Land mined for mineral resources also must be preserved. On both public and private lands, mined land must be restored according to government regulations. **Reclamation** is the process of restoring land disturbed by mining. Mined areas can be reshaped, covered with soil, and then replanted with trees and other vegetation.

Land used for farming and grazing can be managed to conserve soil and improve crop yield. Farmers can leave crop stalks after harvesting to protect soil from erosion. They also can use organic farming techniques that do not use synthetic fertilizers.

Positive Actions Governments, societies, and individuals can work together to reduce the impact of human activities on land resources. For example, many cities use green spaces to create natural environments in urban settings. Green spaces are areas that are left undeveloped or lightly developed. They include parks within cities and forests around suburbs. Protected forests and parks, such as Yellowstone National Park, shown to the right, are important habitats for wildlife.

THREE-DIMENSIONAL THINKING

Complete the chart by filling in the **effects** of each positive action.

Positive Action	Effects
Governments preserve land.	
Logging company reforests an area.	
City creates urban park.	
Girl recycles paper.	
Farmer composts food scraps.	

GO ONLINE for additional opportunities to explore!

Investigate how Earth's land resources are managed by performing one of the following activities.

☐ **Read** about mitigation practices in the **Scientific Text** *Dwindling Land Resources and Urban Farming.*

OR

☐ **Investigate** how to use land responsibly in the **Lab** *Managing Land Resources.*

COLLECT EVIDENCE

How do humans monitor and minimize the impacts of land resources? Record your evidence (B) in the chart at the beginning of the lesson.

LESSON 1

Review

Summarize It!

1. **Record** some of the negative and positive impacts that humans have on the land.

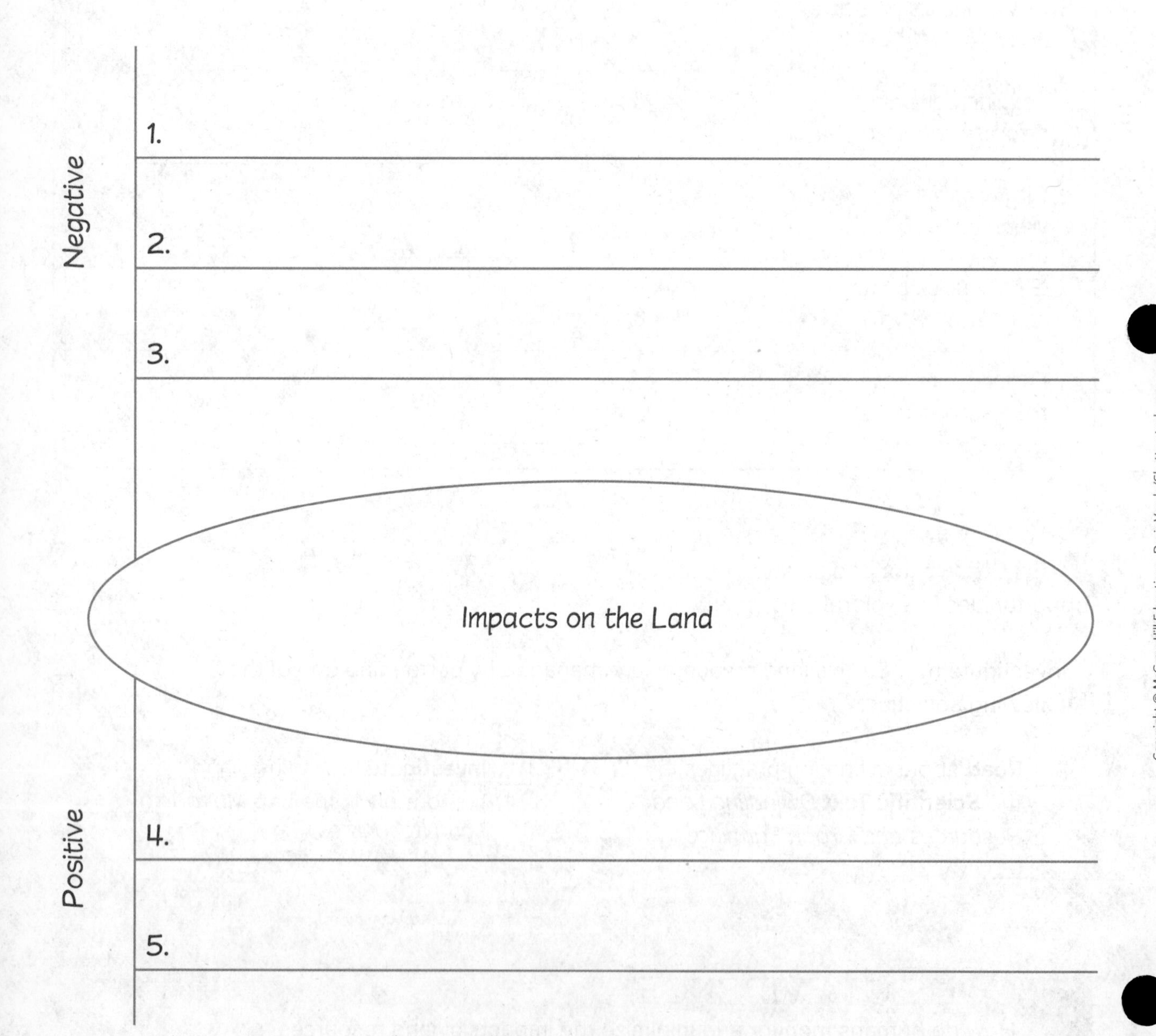

Three-Dimensional Thinking

Study the nitrogen cycle shown in the figure below. Nitrogen is an element that cycles naturally through ecosystems. Living things use nitrogen to make proteins. When these living things die and decompose or produce waste, they release nitrogen into the soil or the atmosphere. Scientists estimate that human activities have doubled the amount of nitrogen cycling through ecosystems.

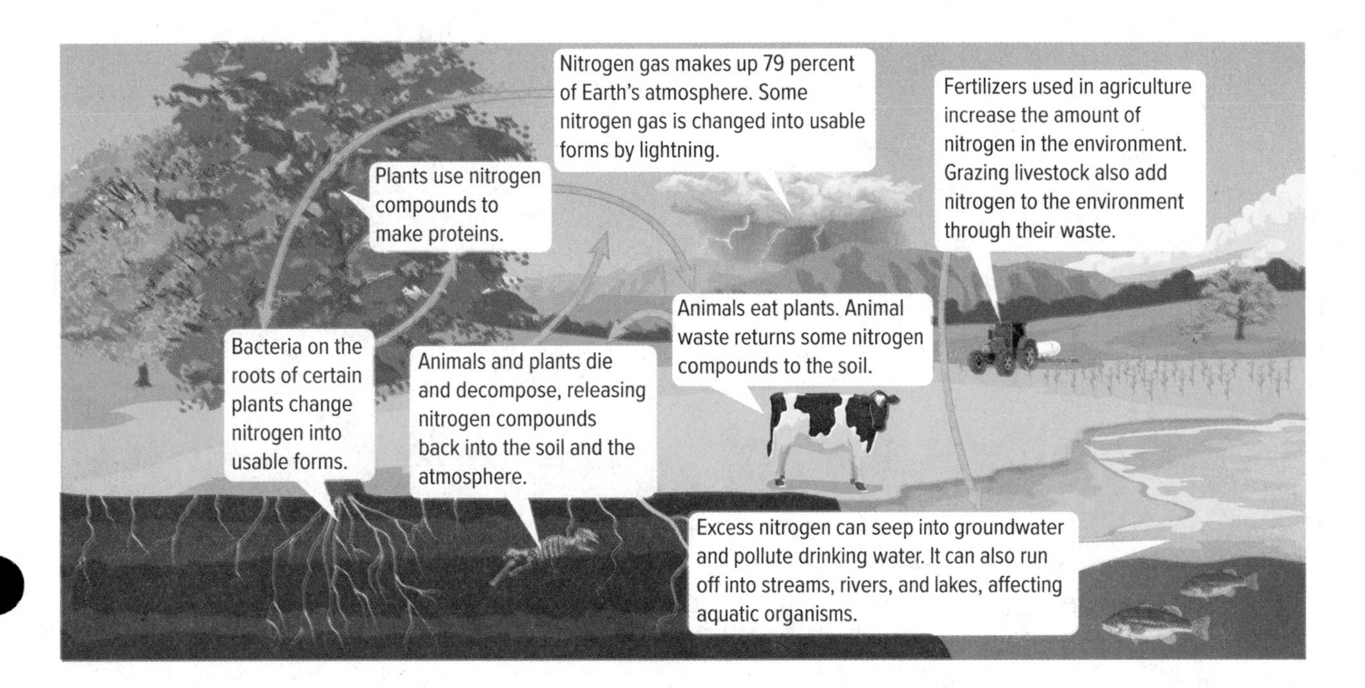

2. How does the use of fertilizers affect the environment?

A Fertilizers decrease the amount of nitrogen in the environment. A decrease in nitrogen can cause an increase in lightning and storms.

B Fertilizers increase the amount of nitrogen in the environment. Excess nitrogen can pollute groundwater and surface water.

C Fertilizers decrease the amount of nitrogen in the environment. This affects the rate at which plants and animals decompose.

D Fertilizers increase the amount of nitrogen in the environment. An increase in nitrogen disrupts plant processes.

Real-World Connection

3. **Describe** Where are three locations in your daily life where paper or cardboard could be recycled? If there is no provision for recycling in any of those locations, what could you do to improve the situation?

Still have questions?
Go online to check your understanding about human impact on land.

REVISIT Do you still agree with the student you chose at the beginning of the lesson? Return to the Science Probe at the beginning of the lesson. Explain why you agree or disagree with that student now.

EXPLAIN THE PHENOMENON

Revisit your claim about how humans impact Earth's land resources. Review the evidence you collected. Explain how your evidence supports your claim.

START PLANNING

STEM Module Project
Engineering Challenge

Now that you have learned about how humans impact Earth's land resources, go to your Module Project to begin organizing information that will help you design a solution to monitor and minimize human impact on the environment. Keep in mind that you want to explain the impact that human activities have on land resources.

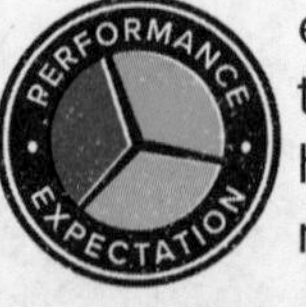

LESSON 2 LAUNCH

Warning: No Swimming

Three friends took a trip to the beach. Upon arrival they saw a sign warning them not to swim because contact with water containing runoff may cause illness. Place a checkmark next to the things you think could be causing the runoff pollution in the ocean.

_____ Lawn fertilizer

_____ Trash

_____ Dirty mop water

_____ Pet waste

_____ Rainwater

_____ Car fuel and oil

Explain your thinking. Describe your reasoning for choosing if something was a cause of pollution to the ocean.

____

____

____

____

____

____

____

You will revisit your response to the Science Probe at the end of the lesson.

LESSON 2

Impact on Water

How can satellites help monitor human impact on Earth's water?

GO ONLINE
Watch the video *Ocean Color Time Series* to see this phenomenon in action.

The National Aeronautics and Space Administration (NASA) used the Landsat 8 satellite to capture this picture of algal blooms in Lake St. Clair near Detroit, Michigan. The video *Ocean Color Time Series* was captured with NASA's *Aqua* satellite. These are just a few examples of how technology is used to monitor human impact on water.

Brainstorm a list of five activities that you think might affect Earth's water quality or supply. How might these activities also impact the biosphere? Next to each activity, list the potential negative effects as well as a potential counter activity or technology that might monitor or minimize the impact.

EXPLAIN
THE PHENOMENON

Are you starting to get some ideas about how human activities can alter Earth's systems and how we might mitigate these effects? Use your observations to make a claim about the ways in which humans can negatively—and positively—impact Earth's water.

CLAIM

It is important to monitor Earth's water, such as with a satellite, because...

COLLECT EVIDENCE as you work through the lesson. Then return to these pages to record your evidence.

EVIDENCE

A. What evidence have you discovered to explain the impacts of water usage?

B. What evidence have you discovered to explain the causes and effects of water pollution?

MORE EVIDENCE

C. What evidence have you discovered to explain how humans can monitor and minimize these impacts?

When you are finished with the lesson, review your evidence. If necessary, based on the evidence, revise your claim.

REVISED CLAIM

It is important to monitor Earth's water, such as with a satellite, because...

Finally, explain your reasoning for how and why your evidence supports your claim.

REASONING

The evidence I collected supports my claim because...

How do humans use water?

Most of Earth's surface is covered with water, and living things on Earth are made mostly of water. Neither the largest whale nor the smallest alga can live without this important resource. Like other organisms, humans need water to survive. However, humans tend to use more water than we need. Let's explore how much water can be wasted with something as small as a leaky faucet. Act as an environmental consultant and analyze water waste.

Drip Drop

Safety 

Materials

leaking faucet

beaker

stopwatch

graduated cylinder

Procedure

1. Read and complete a lab safety form.
2. Catch the water from a leaking faucet in a beaker. Time the collection for 1 minute with a stopwatch.
3. Use a graduated cylinder to measure the amount of water lost.
4. Follow your teacher's instructions for proper cleanup.

Data and Observations

5. Record the amount of water, in milliliters per minute, that leaked from the faucet.

6. Fill in the table to show how much water would leak over time.

Time	1 day	1 week	1 month	1 year
Amount of water from leak (mL)				

Analyze and Conclude

7. Construct a graph of your data. Label the axes and write a title for your graph. In your Science Notebook, explain what the graph illustrates.

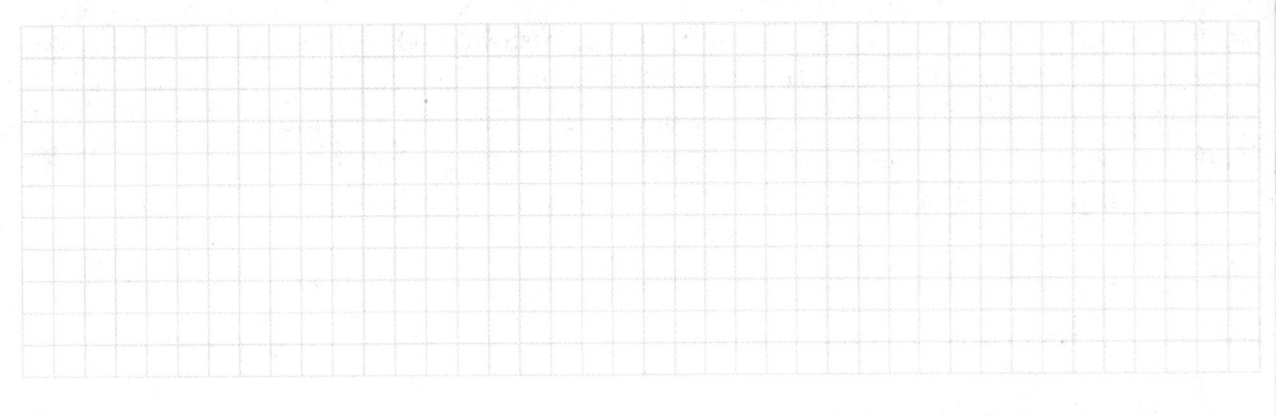

8. As an environmental consultant, you are asked to write a report detailing your findings. What information and recommendations would your report contain about water waste?

Human Water Usage As the human population increases, so does its impact on water usage. Humans also use water in ways that other organisms do not. People wash cars, do laundry, and use water for agriculture, recreation, and transportation. Household activities, however, make up only a small part of human water use. As shown to the right, most water in the United States is used by power plants. The water is used to generate electricity and to cool equipment. The use of water as a resource impacts the environment in many different ways.

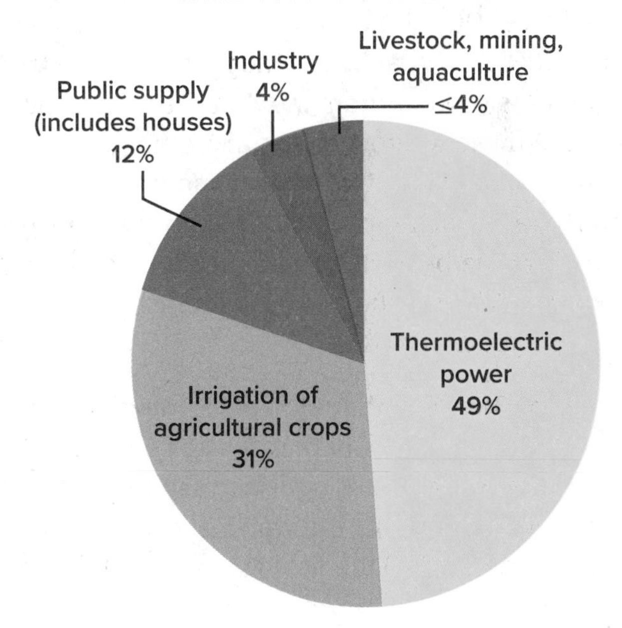

Want more information?
Go online to read more about human impacts on water.

FOLDABLES
Go to the Foldables® library to make a Foldable® that will help you take notes while reading this lesson.

What impacts do humans have on water distribution and availability?

Most of the human activities that require water, require freshwater. Of all the water on Earth, freshwater makes up only 3 percent. Traditional sources of freshwater include ground water, lakes, rivers, and other surface water. As the human population increases, some natural systems are pushed beyond their limits. Let's examine one resource that couldn't keep up with human demand.

INVESTIGATION

Case Study: The Aral Sea

Read the text below then respond to the prompt.

ENVIRONMENTAL Connection Prior to the 1970s, the Aral Sea, which is located in central Asia, was the fourth largest lake in the world. The lake was supplied with freshwater from two major rivers, the Amu Darya and Syr Darya, and located in a basin where the only water loss was due to evaporation. Local economies were supported by the fisheries in the lake.

The Aral Sea is located in Kazakhstan and Uzbekistan (formerly the Soviet Union).

In the 1960s, the Soviet government, which had control of the Aral Sea at the time, diverted large amounts of water from the Amu Darya and Syr Darya rivers to grow cotton and other crops. The engineers behind this project knew this withdrawal would cause the sea to recede. They thought the benefits of an agricultural economy would outweigh the losses to the fishing industry supported by the lake.

Within 10 years, the Aral Sea was rapidly losing water as evaporation exceeded the amount of water still flowing into the lake. The shoreline began to shrink, causing fisherman to travel farther to the water. The lake's salinity, or how salty the water is, greatly increased. This destroyed the fisheries. By 1980 the fishing industry collapsed.

Not only did the lake dry up, but the two rivers carried fertilizer downstream. The salt left behind from evaporation mixed with these chemicals creating a layer of contaminated salt on the lake bed. This salt was carried by the wind and created less productive crop lands and human health problems.

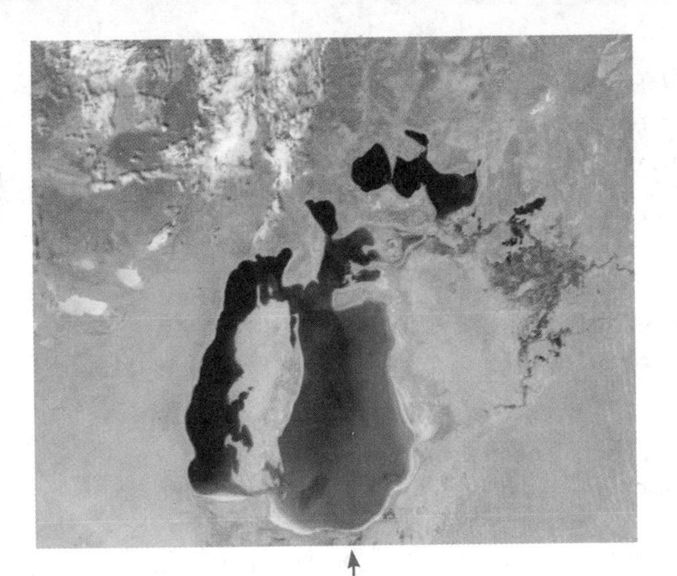

This is the Aral Sea in 2000.

Here is what the Aral Sea looked like in 2017.

Ultimately, the loss of the Aral Sea altered the regional climate and led to many health and economic problems. The human modifications to this natural system produced severe environmental consequences.

WRITING Connection In your Science Notebook, write an essay explaining how the expansion and operation of human communities influenced the Aral Sea natural system.

Changing the Flow of Surface Water People worldwide depend on sources of freshwater for their water supplies. Another method to obtain water is through dams. Streams and rivers are often dammed to create reservoirs that store water.

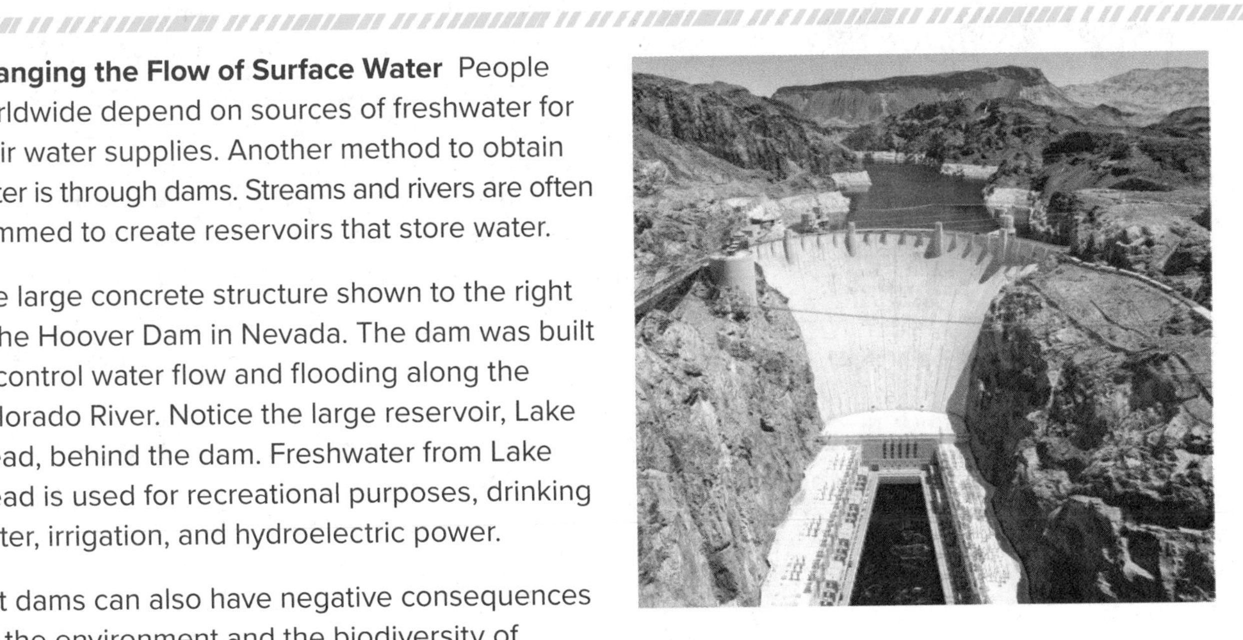

The large concrete structure shown to the right is the Hoover Dam in Nevada. The dam was built to control water flow and flooding along the Colorado River. Notice the large reservoir, Lake Mead, behind the dam. Freshwater from Lake Mead is used for recreational purposes, drinking water, irrigation, and hydroelectric power.

But dams can also have negative consequences on the environment and the biodiversity of ecosystems around the river. Dams can increase the rate of erosion along the banks of the streams. They also act as a geographic barrier for migratory fish. Because of dams, some rivers, such as the Colorado River, are nearly dry before they reach the ocean.

Read a Scientific Text

While dams can have positive effects for humans and other organisms living upstream, dams have many negative effects, particularly on ecosystems downstream of the dam.

CLOSE READING

Inspect

Read the passage *Environmental Issues Associated with Dams.*

Find Evidence

Reread the passage. Highlight the positive effects of dams. Underline the negative effects of dams.

Make Connections

Communicate With your partner, engage in a debate over the construction of a new dam. One of you must support the project, while the other must oppose it. Be sure to use evidence to support your argument.

Environmental Issues Associated with Dams

Dams hold the promise of a nearly constant supply of electrical power with no associated emissions of greenhouse gases or toxic contaminants. They have some advantages compared to other ways of generating electricity, such as the ability to rapidly change the amount of electricity being generated, which reduces the amount of energy lost.

Unfortunately, dams have negative effects on the environment. When filling the reservoir any land located in the area where the reservoir will be, such as farm land, houses, or cities can be destroyed. The flooding will also destroy all the existing vegetation and animal habitat in that area.

Dams also interfere with the natural river dynamics, changing the natural flow patterns such as creating a more consistent flow. A dam traps sediment carried by the rivers and streams flowing into the reservoir, and the reservoir will eventually fill up with sediment. This blockage deprives the downstream river of sediment and associated nutrients. This can cause erosion and drastically change the downstream habitat, as clear, cold water from the depths of the reservoir replaces the warmer, muddy water that flowed down the river before the construction of the dam.

Dams are also lethal for migratory fish, such as salmon. Adult fish are blocked from migrating to upstream spawning areas. Juvenile fish die if they go through hydroelectric turbines.

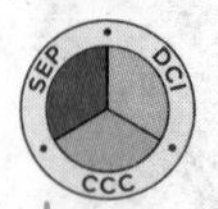

THREE-DIMENSIONAL LEARNING

Write a short statement to present to your city council supporting or not supporting the construction of a new dam in your community. What **changes** would a dam bring to your local ecosystem? **Explain** your reasoning.

Impacts on Groundwater As you read, human activity has many impacts on the distribution and availability of surface water. Humans have an impact on groundwater as well. The water beneath Earth's surface is much more plentiful than the freshwater in lakes and streams. Recall that groundwater makes up the majority of Earth's freshwater. Groundwater is an important source of water for many streams, lakes, and wetlands. Some plant species absorb groundwater through long roots that grow deep underground. People in many areas of the world rely on groundwater for their water supply. In the United States, about 20 percent of the water people use daily comes from groundwater. People often bring groundwater to Earth's surface by drilling wells. Wells are usually drilled into an **aquifer**—an area of permeable sediment or rock that holds significant amounts of water. Groundwater then flows into the well from the aquifer and is pumped to the surface.

INVESTIGATION

Sinkholes

GO ONLINE Watch the animation *Groundwater and the Formation of Sinkholes* and answer the following question.

Explain the cause and effect of overdrawing ground water.

Groundwater Precipitation helps replace groundwater drawn out of wells. During a drought, less groundwater is replaced, so the water level in a well drops. The same thing happens if water is removed from a well faster than it is replaced. If the water level drops too low, a well runs dry. The water in an aquifer supports the rocks and soil above it. In some parts of the world, water is being removed from aquifers faster than it can be replaced. This creates empty space underground. The empty space underground cannot support the weight of the overlying rock and soil. Sinkholes form where the ground collapses due to lack of sufficient support from below.

COLLECT EVIDENCE

What evidence have you discovered to explain the impacts of water usage? Record your evidence (A) in the chart at the beginning of the lesson.

A Closer Look: San Joaquin Valley

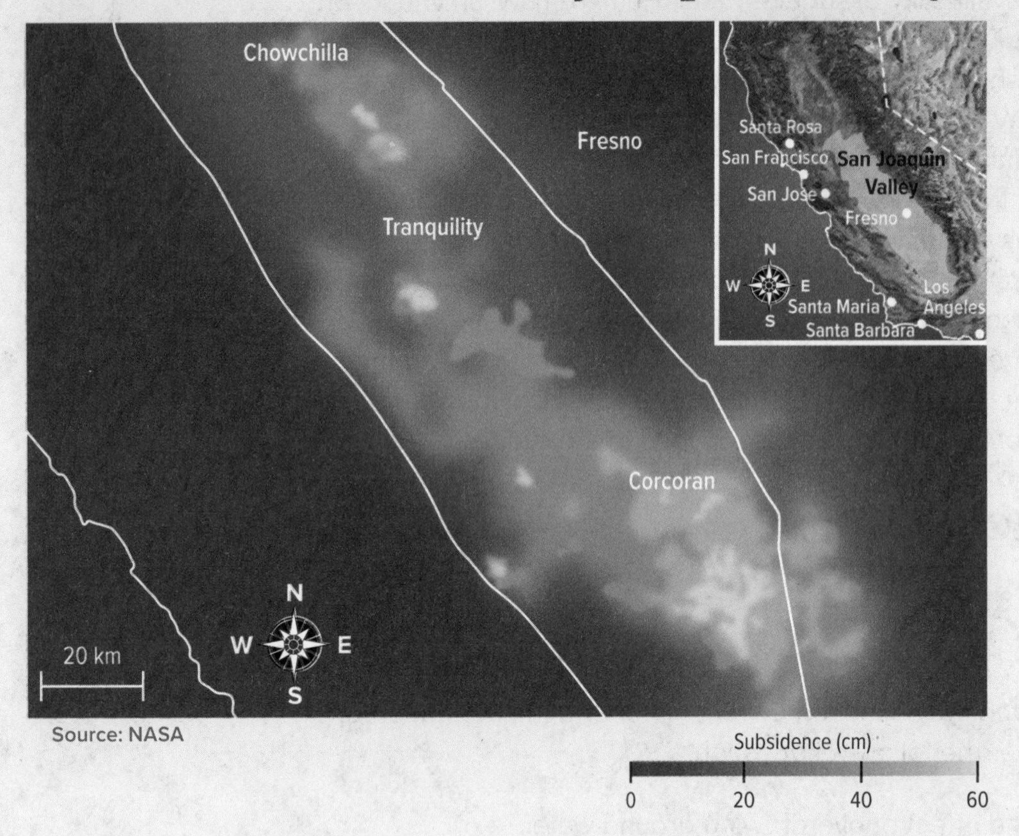

Source: NASA

The San Joaquin valley is one of the most productive agricultural regions in the nation. All of the agriculture in the area creates a high demand for water. Farmers have been withdrawing groundwater from the valley since the 1920s. By the 1970s, some areas had subsidence, or sinking, of up to 8.5 m!

Think about what happens when you leave a sponge out to dry. It shrinks! This is similar to what is happening in the San Joaquin Valley. As water is taken out of the ground, the pores are left empty. The pressure from everything above ground causes the soil to become compacted.

The valley is still subsiding with sinking increasing during periods of drought. The subsidence negatively impacts California water supplies, bridges, roads, groundwater wells, permanent groundwater storage, and more. California state and federal water agencies have spent an estimated $100 million on related repairs to the San Joaquin Valley since the 1960s.

It's Your Turn!

MATH Connection Investigate the rate at which land is subsiding in the San Joaquin Valley. If the valley is sinking at an average rate of 30 cm a year, how much has been lost over the past five years? What reduction do you predict in the next ten years? Create a public service announcement to present your data to neighborhoods in the area.

How do humans pollute Earth's water?

Groundwater pollution is a serious problem. It becomes more serious as the human population grows and the use of this resource increases. When water seeps into the ground, it can carry with it many pollutants. Let's explore how water pollution travels!

LAB Pollution in Motion

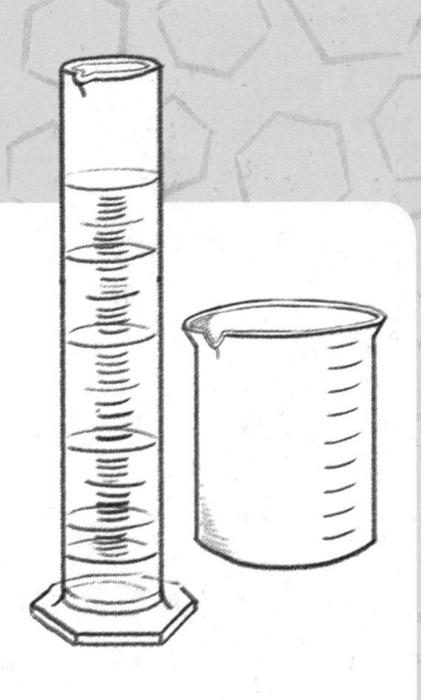

Safety 

Materials

clear plastic container
aquarium gravel
water
pump (from liquid soap or similar)
food coloring
5 cm x 5 cm piece of cloth
small rubber band
graduated cylinder
30 mL of salt water
beaker

Procedure

1. Read and complete a lab safety form.
2. Add clean gravel to a clear plastic container to a depth of 1 cm.
3. Pour water into the container until it is just below the top of the gravel.
4. Cover the bottom end of the pump with cloth and secure it with a rubber band. Hold the pump in the gravel near one end of the container.
5. Use a few drops of food coloring to dye 30 mL of salt water.
6. Pour the salt water on top of the gravel at the end of the container away from the pump.
7. Observe the salt water's movement.
8. Draw the experiment in the Data and Observations section on the next page. Illustrate what you think will happen to the salt water once you begin pumping.
9. Begin pumping the water into the empty beaker. Observe the water.
10. Follow your teacher's instructions for proper cleanup.

Data and Observations

Analyze and Conclude

11. How did the pump affect the movement of water?

12. How did pumping water affect the movement of pollutants?

13. How could an increasing population affect the movement of pollutants and the quality of water in drinking wells?

Water Pollution Human activity often increases the amount of pollutants in Earth's water. While humans are the cause of this pollution, many types of pollution have outcomes that are not beneficial to humans such as dead zones and polluted drinking water. Examine the figures below to see different sources of water pollution.

SCIENCE & SOCIETY

AMERICAN MUSEUM OF NATURAL HISTORY

Down the Drain

Flushing Chemicals into Fish Habitats

Anytime you flush a toilet or run a faucet, you produce wastewater. When chemicals are added to wastewater, they sometimes can end up in lakes and rivers, threatening wildlife. In recent years, scientists have discovered that when the chemical estrogen is in wastewater, it harms wild fish populations.

Estrogen is a chemical responsible for sexual development and reproduction in female vertebrates. It is produced naturally in the body, but it also is in medications that many women take. Like other chemicals in drugs, estrogen is released in urine and ends up in wastewater.

If even a small amount of estrogen gets into waterways, it can have a huge impact on local fishes. The chemical disrupts the organs that enable them to reproduce. Most fishes are not born male or female. Instead, environmental factors, such as temperature, food availability, and social interactions, determine which sex organs they develop. In a healthy habitat, some fishes produce eggs and others produce the sperm to fertilize eggs. However, increased levels of estrogen in the water can affect the expression of traits in fish. Data collected in the field and in the laboratory show that estrogen affects how fish develop. When exposed to estrogen, males produce fewer or no sperm and some even produce eggs. Many females stop producing eggs. Without healthy males and females to reproduce, these fish populations drop.

Lake Experiment

To understand estrogen's effects on wild fish, scientists released small amounts of the chemical into a lake in Ontario, Canada. They observed drastic changes in the fathead minnows that had been thriving in the lake. The females produced fewer eggs. The males produced fewer sperm or began to develop eggs. After 3 years, the minnow population had nearly disappeared. Once scientists stopped adding estrogen, the fish population began to recover.

It's Your Turn

Research how pesticides or other chemicals are affecting a particular animal species. Include the name of the chemical, how it is used, and how it affects the animal. Create a warning label for the product to warn consumers of the effects.

Ocean Pollution Have you ever seen a photograph of a shorebird or seal covered with oil? Spills from oil tankers harm wildlife. They also harm the ocean. Any harm to the physical, chemical, or biological health of the ocean ecosystem is ocean pollution. Sometimes ocean pollution comes from a natural source, such as a volcanic eruption. More often, human activities cause ocean pollution.

The figure to the right shows the proportion of different sources of ocean pollution caused by humans. Notice that only 13 percent of this pollution comes from shipping or offshore mining activity. The rest comes from land. Land-based pollution includes garbage, hazardous waste, and fertilizers. Airborne pollution that originates on land, such as emissions from power plants or cars, is also included in this category. So is trash dumped directly into the oceans. Have you ever seen a turtle tangled in plastic? How did this happen?

Ocean Pollution Sources

Runoff from land 44%
Airborne pollutants that originate on land 33%
Spills from shipping 12%
Dumping trash directly into the ocean 10%
Offshore mining and drilling for resources 1%

LAB Waves of Waste

Safety

Materials

bowl

water

objects

Procedure

1. Read and complete a lab safety form.
2. Half-fill a large bowl with water.
3. Sprinkle the objects you have collected into the water.
4. Gently swirl the water in the bowl until the water moves at a constant speed. Try not to create a whirl pool.
5. Follow your teacher's instructions for proper clean up.

Analyze and Conclude

6. What happened to the objects in the bowl?

7. What do you think happens to litter in the ocean?

8. What do you think can be done to prevent ocean pollution?

LIFE SCIENCE Connection Ocean pollution, like the objects you used in the Lab *Waves of Waste,* has both immediate effects and long-term effects on marine systems. *Marine* refers to anything related to the oceans. Chemical waste can be poisonous to marine organisms. Fish and other organisms absorb the poison and pass it up the food chain. A large oil spill can harm marine life. So can solid waste, excess sediments, and excess nutrients. Let's explore how solid waste such as garbage affects the oceans.

INVESTIGATION

Ocean Garbage

GO ONLINE Watch the video *Rubbish* then answer the following question.

What is the cause and effect of the Great Pacific Garbage Patch? How could we reduce the size of the patch?

Solid Waste Trash, including plastic bottles and bags, glass, and foam containers, causes problems for marine organisms. Many birds, fish, and other animals become entangled in plastic or mistake it for food. Plastic breaks up into small pieces, but it does not degrade easily. Some of it becomes trapped in circular currents, or gyres. The North Pacific Gyre has collected so much plastic and other debris that a portion of it has been dubbed "The Great Pacific Garbage Patch." The patch is thought to be twice the size of Texas. About 80% of the debris originates from land-based activities in North America and Asia.

Excess Sediments Large amounts of land-based sediment wash into oceans, as shown to the right. Some erosion is natural. However, some is caused by humans, who cut down trees near rivers and ocean shorelines. Without the roots of trees and other vegetation to hold sediments in place, the sediments more readily erode. Excess sediments can clog the filtering structures of marine filter feeders, such as clams and sponges. Excess sediments also can block light from reaching its normal depth. Organisms that use light for photosynthesis could die.

Excess Nutrients Algae need nutrients such as nitrogen and phosphorus to survive and grow. For this reason pollution from excess nutrients is beneficial to algae. However, too many nutrients can cause an explosion in algal populations. An algal bloom occurs when algae grow and reproduce in large numbers. Algal blooms also can cause water to appear red, green, brown, or even glow at night. Nitrates and phosphates can be abundant in agricultural runoff as well as coastal upwelling zones. Many scientists have found that a major source of excess nitrates and phosphates is from land-based fertilizers that wash into oceans.

These algae are bioluminescent—they glow!

COLLECT EVIDENCE

What are the causes and effects of water pollution? Record your evidence (B) in the chart at the beginning of the lesson.

Dead Zones

GREEN SCIENCE

What causes lifeless areas in the ocean?

For thousands of years, people have lived on coasts, making a living by shipping goods or by fishing. Today, fisheries in the Gulf of Mexico provide jobs for thousands of people and food for millions more. Although humans and other organisms depend on the ocean, human activities can harm marine ecosystems. Scientists have been tracking dead zones in the ocean for several decades. They hypothesize that these zones are a result of human activities on land.

A large dead zone in the Gulf of Mexico forms every year when runoff from spring and summer rain in the Midwest drains into the Mississippi River. The runoff contains nitrogen and phosphorous from fertilizer, animal waste, and sewage from farms and cities. This nutrient-rich water flows into the gulf. Algae feed on excess nutrients and multiply rapidly, creating an algal bloom. The results of the algal bloom are shown below.

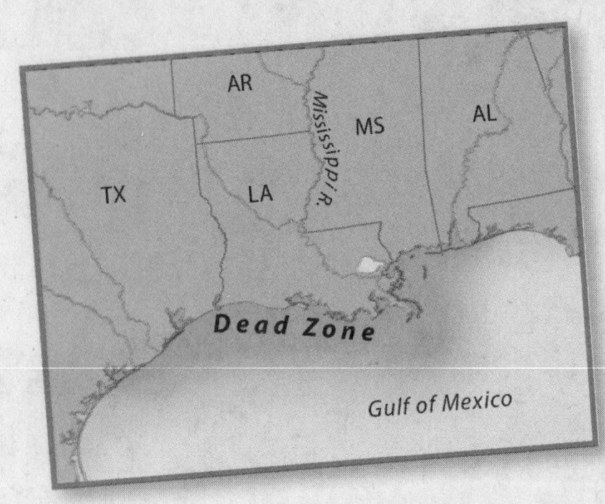

Some simple changes in human activity can help prevent dead zones. People upstream from the Gulf can decrease the use of fertilizer and apply it at times when it is less likely to be carried away by runoff. Picking up or containing animal waste can help, too. Also, people can modernize and improve septic and sewage systems. How do we know these steps would work? Using them has already restored life to dead zones in the Great Lakes!

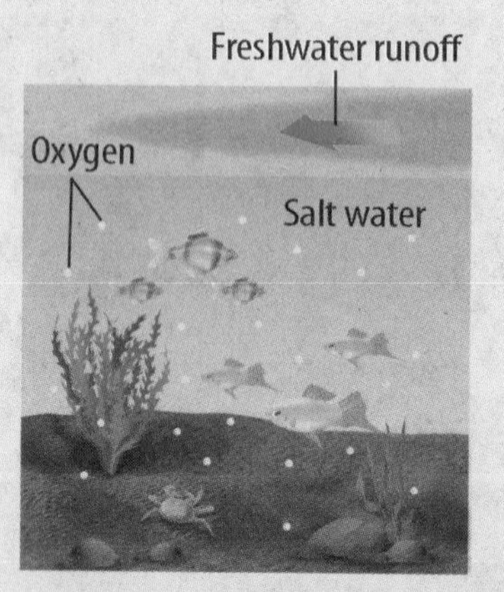

① **River water containing nitrogen and phosphorous flows into the Gulf of Mexico.**

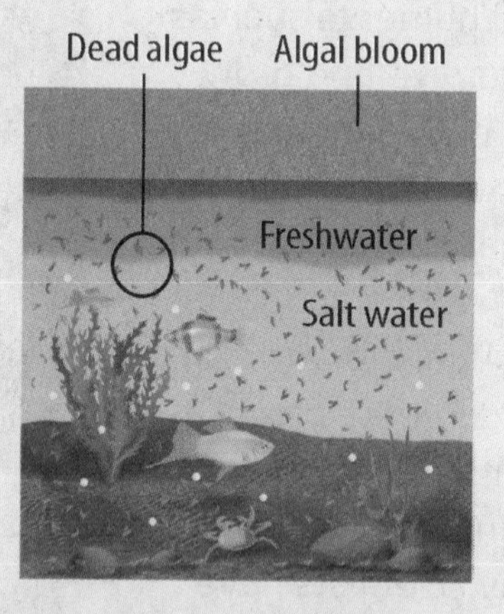

② **After the algal bloom, dead algae sink to the ocean floor.**

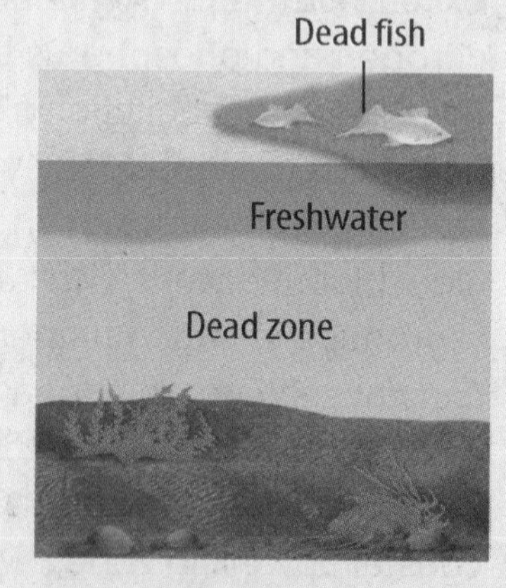

③ **Decomposing algae deplete the water's oxygen, killing other organisms.**

It's Your Turn

Research and Report Earth's oceans contain about 150 dead zones. Choose three. Plot them on a map and write an article for a website about what causes each dead zone.

How can we monitor and minimize human impact on water?

All life on Earth needs water, but humans are the only organisms that unnecessarily use and pollute water. When humans have a negative impact on natural resources such as water, it does not only affect us, it affects all life on Earth. What are some ways to reduce the impact of the human population on water resources?

These pelicans are covered in oil from an oil spill.

 ENGINEERING INVESTIGATION

Solutions for Pollution

You are part of a team of water-pollution specialists. For your job, you examine various water-pollution problems. Your boss has requested that you research and evaluate ways to solve water-pollution problems. She suggests that you and your team research a current problem and be prepared to determine the best possible solutions for your clients.

1. Review the types of water pollution you have learned about.
2. Choose one source of pollution you would like to focus your research on.
3. Use print and digital sources to research possible solutions for this form of pollution. Don't forget to make sure the sources are credible!
4. Use the information from your research to fill in the table on the next page.
5. Once you have completed your research, compare and contrast your solution with your classmates' solutions.
6. Recommend the best overall solution for your type of water pollution in a concise report to your boss in your Science Notebook. Support your decision with the evidence you have gathered.

Solution Evaluation	
How does the solution treat the pollution?	
Will the solution fully restore clean water? Why or why not?	
Is the solution eco-friendly? How does it protect humans and the environment?	
How will it help avoid future pollution?	
What materials or equipment may be needed to build/ maintain the solution?	
How have similar problems been solved in the past?	
What impacts will this solution have on society?	
How much time is required to build the solution?	

Management Solutions Legislation is an effective way to reduce water pollution. The U.S. Clean Water Act legislates the reduction of water pollution. This act established the basic procedures for regulating wastewater and setting pollution controls. While the Clean Water Act has greatly helped reduce pollution, water pollution in the U.S. still exists.

The Safe Drinking Water Act legislates the protection of drinking water supplies. By reducing pollution, these laws help ensure that all living things have access to clean water.

In 1969, the Cuyahoga River caught on fire, prompting the Clean Water Act.

What You Can Do An easy way for you to make an impact on water conservation is to use less. This could be as simple as taking shorter showers or turning off the water while you brush your teeth. You can have an even bigger impact by getting others to conserve with you.

To reduce pollution, you could recycle. By recycling your plastic drinking bottles, you can ensure they won't end up floating in the ocean. Another approach is to stop it before it starts—by reducing! If you reduce your plastic use there will be less in the garbage or even the recycling.

You also can help reduce water pollution by properly disposing of harmful chemicals so that less pollution runs off into rivers and streams. You can volunteer to help clean up litter from a local stream. You also can conserve water so there is enough of this resource for you and other living things in the future.

COLLECT EVIDENCE

What are ways in which we can monitor or minimize human impact on Earth's water? Record your evidence (C) in the chart at the beginning of the lesson.

LESSON 2

Review

Summarize It!

1. **Record** some of the negative and positive impacts that humans have on Earth's water.

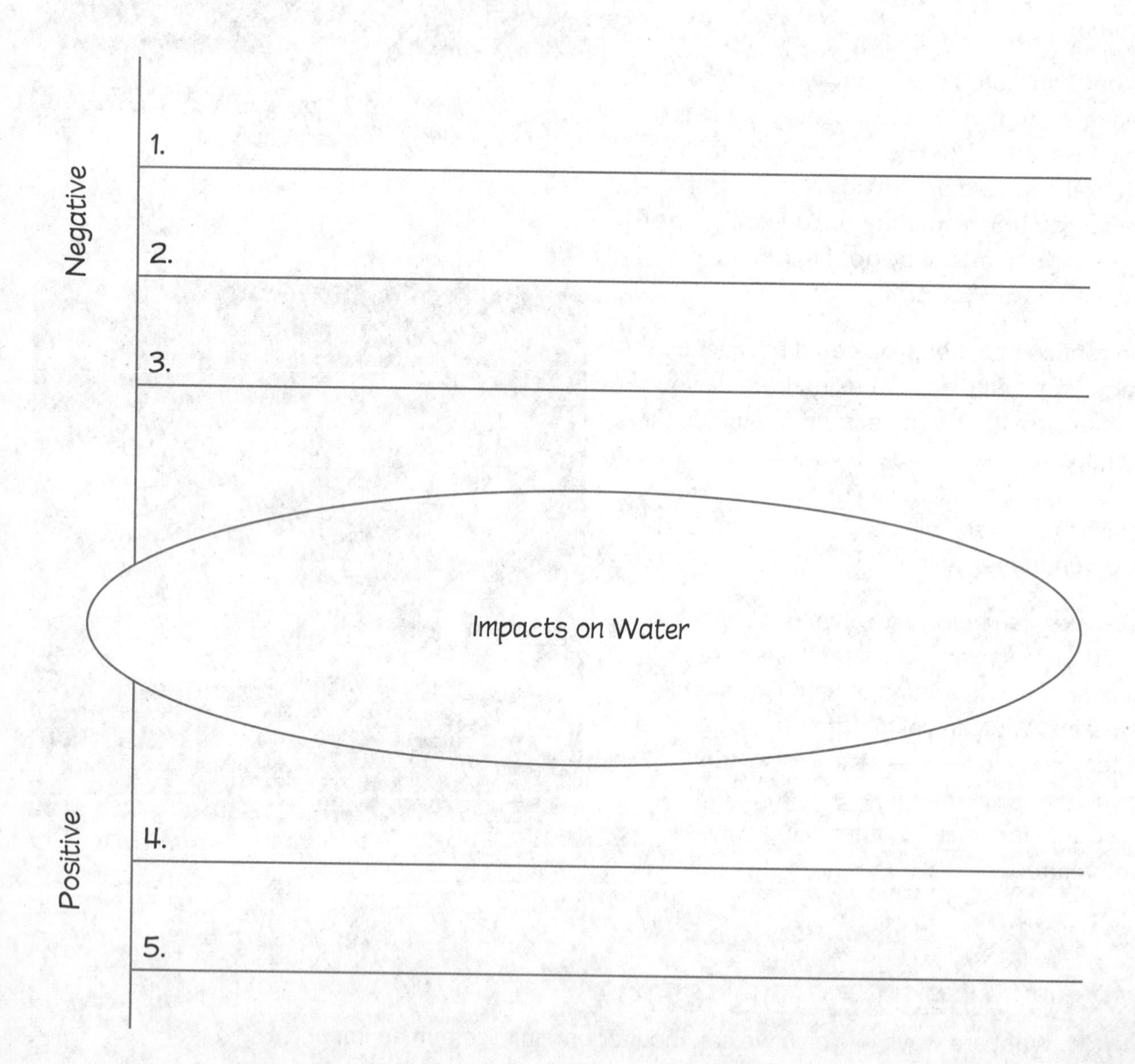

Three-Dimensional Thinking

Read the passage below. Then answer the question that follows.

> Estuaries form where rivers containing freshwater flow into the salty waters of an ocean. The mixture of fresh and salt water stays balanced as long as both the river and ocean tides continue to mix at the river's mouth. Estuaries are usually calm and often contain many food sources. Because of this, many species of fish and other organisms breed and raise their offspring in estuaries. These organisms are adapted to life in brackish estuary waters.

2. **LIFE SCIENCE Connection** A new recreation area is being built upstream from an estuary that is known for its abundance of fish and turtles. A dam will be built across the river and a large lake will form behind it. What effect will the dam have on the organisms living in the estuary?

A Organisms adapted to living only in brackish water will survive.

B Organisms adapted to living in brackish water will move to live in the open ocean.

C Some of the organisms will die because the water will be less salty.

D Some of the organisms will die because the water will be more salty.

The table lists a family's average water usage per day for several activities.

Daily Activity	Water Used
Taking a short shower	95 L
Taking a bath	150 L
Running the dishwasher	38 L
Watering lawn and plants	30 L

3. **MATH Connection** The family wishes to reduce their weekly water usage as much as possible. Here are three proposals for how to do this. Which will save the most water?

A Proposal 1: Five times a week substitute a short shower for bath.

B Proposal 2: Run the dishwasher five times a week instead of seven times a week.

C Proposal 3: Water the plants three times a week instead of five times a week.

Real-World Connection

4. **Predict** Your family is thinking about clearing the trees in your backyard to put in a swimming pool. What impact could this have on a local stream? Think of one solution for this problem.

5. **Describe** two ways you can personally help minimize human impact on Earth's water.

Still have questions?
Go online to check your understanding about how human activities impact Earth's water.

REVISIT SCIENCE PROBES

Do you still agree with your choices at the beginning of the lesson? Return to the Science Probe at the beginning of the lesson. Explain why you agree or disagree your choices now.

Revisit your claim about why it is important to monitor Earth's water. Review the evidence you collected. Explain how your evidence supports your claim.

KEEP PLANNING

STEM Module Project Engineering Challenge

Now that you've learned about how humans impact water quality and supply, go back to your Module Project to continue planning your project. Your goal is to explain why it is important to monitor Earth's water. Keep in mind how NASA uses satellites to monitor water resources.

SCIENCE PROBES

LESSON 3 LAUNCH

What is air pollution?

Three friends were talking about air pollution. This is what they said:

Finnegan:	I think air pollution includes solid particles and gases that are harmful to people and the environment.
Sal:	Gases are invisible and can't harm the environment. Air pollution refers only to particles in the air.
Valeria:	Particles are solid so they can't be part of the atmosphere. I think air pollution only includes harmful gases.

Which friend do you agree with the most? Explain.

You will revisit your response to the Science Probe at the end of the lesson.

LESSON 3

Impact on the Atmosphere

Why are solutions, like vertical forests, important for the health of the atmosphere?

Plant-packed residential high-rises, called vertical forests, are being erected around the world. Vertical forests act like giant air purifiers, absorbing some types of air pollution while producing oxygen.

Brainstorm a list of five activities that you think might cause harm to Earth's atmosphere. Next to each activity, list the potential negative effects as well as a potential counteractivity or technology that might monitor or minimize the impact.

GO ONLINE Watch the video *Above It All* to see this phenomenon in action.

EXPLAIN THE PHENOMENON

Are you starting to get some ideas about activities that might affect Earth's atmosphere and how we might mitigate these effects? Use your observations about the phenomenon to make a claim about the ways in which humans can negatively and positively impact the atmosphere.

CLAIM

Solutions, like vertical forests, are important for Earth's atmosphere because...

COLLECT EVIDENCE as you work through the lesson. Then return to these pages to record your evidence.

EVIDENCE

A. What evidence have you discovered to explain the causes and effects of air pollution?

MORE EVIDENCE

B. What evidence have you discovered to explain how humans can minimize these effects?

When you are finished with the lesson, review your evidence. If necessary, based on the evidence, revise your claim.

REVISED CLAIM

Solutions, like vertical forests, are important for Earth's atmosphere because...

Finally, explain your reasoning for how and why your evidence supports your claim.

REASONING

The evidence I collected supports my claim because...

What is air pollution?

ENVIRONMENTAL Connection As the human population grows, people use more energy to heat and cool homes; to fuel cars, airplanes, and other forms of transportation; and to produce electricity. This energy use contributes to air pollution that affects the composition and viability of the atmosphere. The contamination of air by harmful substances including gases and smoke is called **air pollution.** Types of air pollution include smog, chlorofluorocarbons (CFCs), particulate matter, and acid precipitation. Let's take a closer look at each type of air pollution.

INVESTIGATION

In a Haze

Compare the atmosphere in the two images below.

1. How would you describe the change from one image to the other?

2. What do you think causes this type of change? Write your response in your Science Notebook.

Want more information?
Go online to read more about the impact of human activities on Earth's atmosphere.

FOLDABLES
Go to the Foldables® library to make a Foldable® that will help you take notes while reading this lesson.

Photochemical Smog The brownish haze in the sky pictured in the right image on the previous page is photochemical smog. **Photochemical smog** is caused when nitrogen and carbon compounds in the air react in sunlight. Nitrogen and carbon compounds are released when fossil fuels are burned to provide energy for vehicles and power plants. These compounds react in sunlight and form other substances. One of these substances is ozone. Ozone close to Earth's surface makes air difficult to breathe. It can also damage the tissues of plants and animals.

While ozone in the lower atmosphere is harmful, ozone high in the atmosphere in the ozone layer helps protect living things from the Sun's ultraviolet (UV) radiation. Does air pollution impact the ozone layer?

INVESTIGATION

Oh Ozone

Compare the ozone layer in 1979 (left) to the ozone layer in 2016 (right).

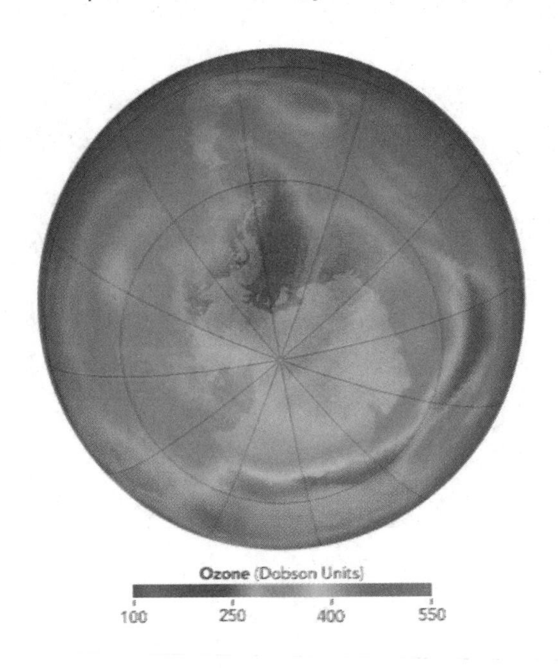

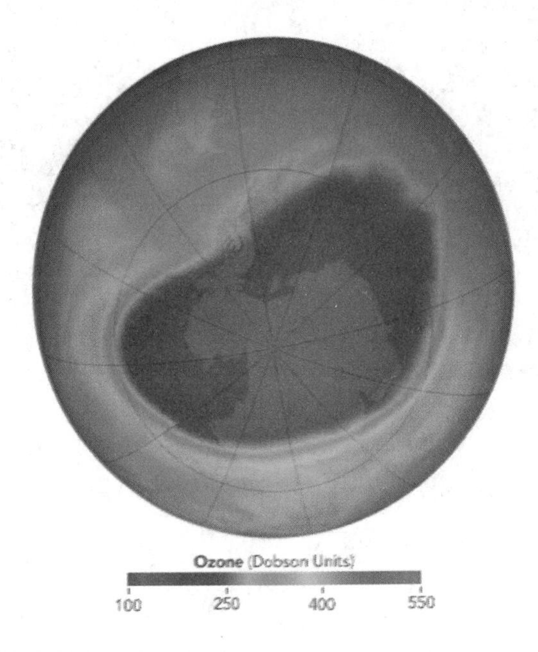

Which image shows a decrease in total ozone? What do you think caused the decrease in ozone?

CFCs In the 1970s, scientists suggested that CFCs could destroy ozone in the upper atmosphere. Studies revealed a thinning of the ozone layer, particularly over Antarctica.

All of the CFCs in the atmosphere are a result of human activity. CFCs are released from products such as old refrigerators and air conditioners, and propellants in aerosol cans. Ozone in the upper atmosphere absorbs harmful UV rays from the Sun. CFCs react with sunlight and destroy ozone molecules. As a result, the ozone layer thins and more UV rays reach Earth's surface. This, in turn, can harm the tissues of plants and animals.

While CFCs indirectly harm organisms, another form of pollution has a direct effect on Earth's biosphere. Let's explore.

INVESTIGATION

As a Matter of Fact

Compare areas along the Yangtze River in China.

How would you describe the atmosphere in the above locations? What might cause the differences you observed?

Particulate Matter The mix of both solid and liquid particles in the air is called **particulate matter.** Solid particles include smoke, dust, and dirt. These particles enter the air from natural processes, such as volcanic eruptions and forest fires. Human activities, such as burning fossil fuels at power plants and in vehicles, also release particulate matter. Inhaling particulates can cause asthma, bronchitis, and lead to heart attacks. It can also interfere with the processes of cellular respiration and photosynthesis in plants.

A Closer Look: Air Pollution and Health Effects

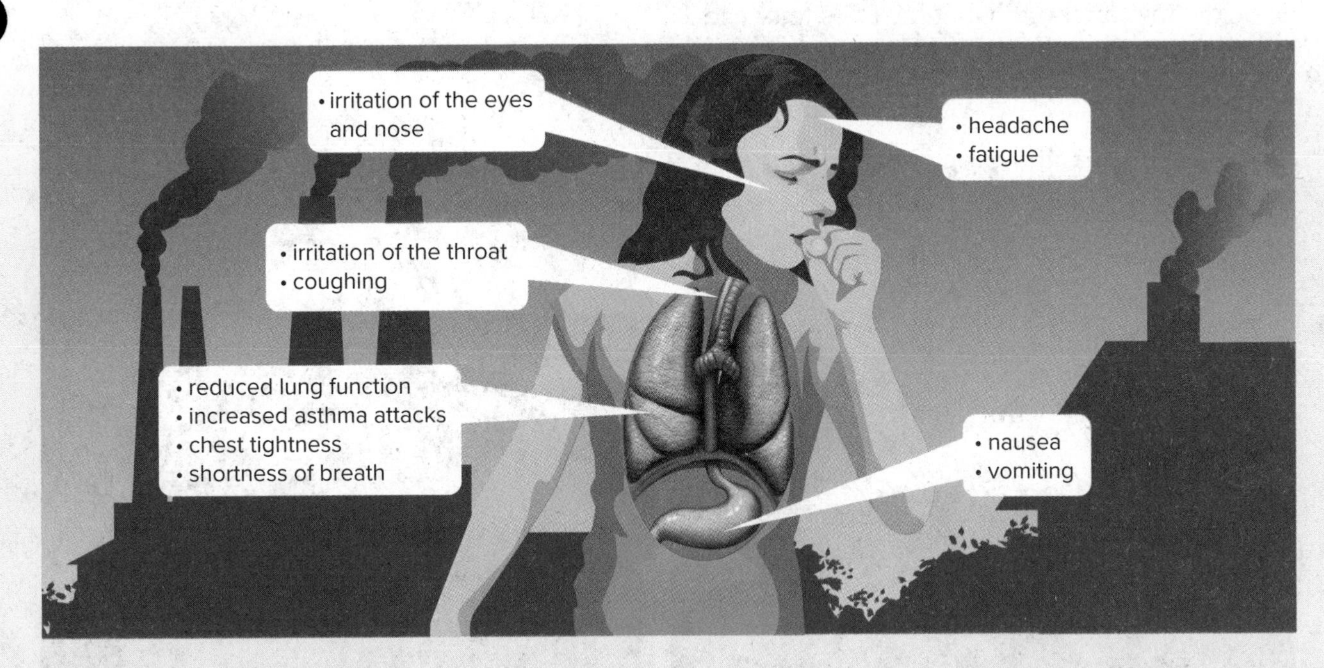

Your body, and the bodies of other animals, uses oxygen in the air to produce some of the energy it needs. But the air you breathe must be clean or it can harm your body. The United Nations estimates that at least 1.3 billion people live in areas where outdoor air is dangerously polluted.

The figure above shows a variety of health problems caused by pollutants in the air. Common diseases related to air pollution include pneumonia, emphysema, and bronchitis. Air pollution can also cause asthma attacks. Asthma is a disorder of the respiratory system in which breathing passageways narrow during an attack, making it hard for a person to breathe. Doctors often prescribe inhalers, like the one shown to the right, to relax and dilate airways to increase airflow. In developed countries, especially in urban areas, asthma cases have steadily increased recently. Today more than 7 percent of children have asthma.

Asthma inhaler

It's Your Turn

HEALTH Connection Conduct a survey to find the rate of occurences of asthma among the population of your classmates, relatives, friends, and/or community members. Interview at least six people in three different generations. Ask each person: "When you were in school, did you know anyone who had asthma?" If the answer is yes, ask "How many people?" Record your data in a table. Combine your data with those of the class. Construct a bar graph of the combined data.

The final major type of air pollution is acid precipitation. What are the effects of acidic rain on the biosphere? Let's take a look.

Damaging Drizzle

Compare areas at the Great Smoky Mountains National Park in Tennessee.

How would you describe the state of the trees in the two photos?

Acid Precipitation The trees you observed above were affected by acid precipitation. **Acid precipitation** is rain or snow that has a lower pH than that of normal rainwater. The pH of normal rainwater is about 5.6.

Acid precipitation forms when gases containing nitrogen and sulfur react with water, oxygen, and other chemicals in the atmosphere. Although volcanoes and marshes add sulfur gases to the atmosphere, burning fossil fuels is a major source of sulfur emissions. Acid rain can pollute soil and harm trees and other plants. When it falls into lakes and rivers, it can harm fish and other organisms. Many living things cannot survive if the pH of water or soil becomes too low.

Now that you understand the causes and effects of acid precipitation, see if your region experiences this type of air pollution. In the following lab, you will test the rainwater around your home.

LAB Close to Home

Safety 

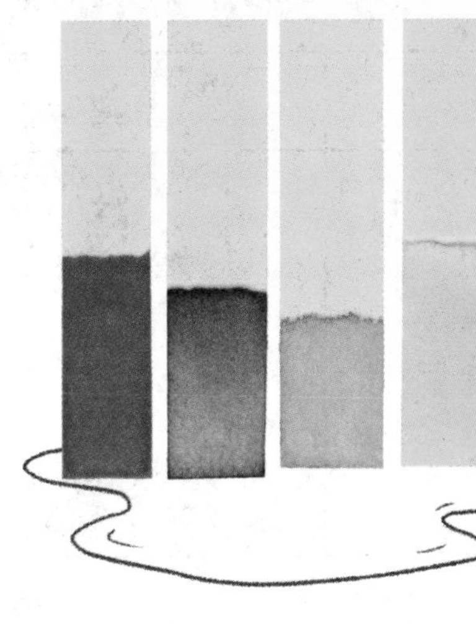

Materials

pH paper

pH color chart

container

Procedure

1. Read and complete a lab safety form.
2. Place a clean container outside in an open area.
3. After a rainstorm, dip a strip of pH paper into the water. Use a pH color chart to determine the pH of the rainwater.

Analyze and Conclude

4. What was the pH of the rainwater that you collected?

5. Does your neighborhood experience acid precipitation? Explain.

6. Knowing the causes of acid rain, what recommendations could you make to reduce this type of air pollution in your region? Write your response in your Science Notebook.

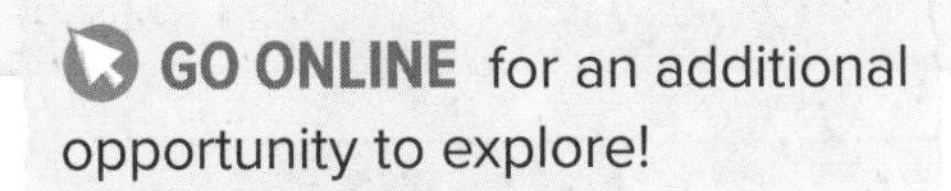

ENVIRONMENTAL Connection Want to learn more about how acid rain can impact the biological diversity and viablilty of natural systems? Then perform the following activity.

- [ ] **Explore** the impacts of acid precipitation in the **Investigation** *pH Tolerance of Aquatic Life.*

THREE-DIMENSIONAL THINKING

Summarize your understanding of the **cause-and-effect** relationships between human activities and/or natural events and the environmental impacts on the atmosphere in the table below.

Type	Causes	Effects
Smog		
CFCs		
Particulate matter		
Acid Precipitation		

COLLECT EVIDENCE

What are the causes and effects of air pollution? Record your evidence (A) in the chart at the beginning of the lesson.

How can we protect the atmosphere?

Preserving the quality of Earth's atmosphere requires the cooperation of government officials, scientists, and the public. Countries around the world are working together to reduce air pollution. For example, 197 countries, including the United States, have signed an international treaty called the Montreal Protocol to phase out the use of CFCs. In the United States, the Clean Air Act sets limits on the amount of certain pollutants that can be released into the air. Some states, such as California, have established their own emissions standards for motor vehicles. At least twelve other states have adopted or plan to adopt standards similar to California's in an attempt to reduce air pollution. Do initiatives such as these work? Let's investigate.

INVESTIGATION

International and National Initiatives

Examine the graph showing global CFC levels below.

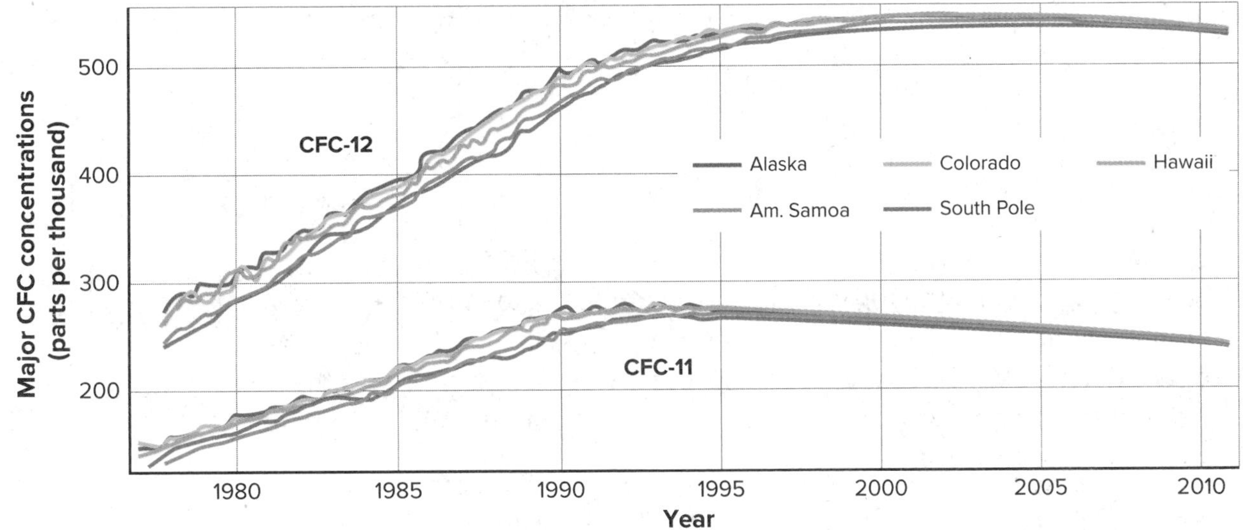

Source: National Oceanic and Atmospheric Administration

1. What trends or patterns do you notice in the data?

Now, take a look at the graph showing atmospheric levels of several types of air pollutants in the United States.

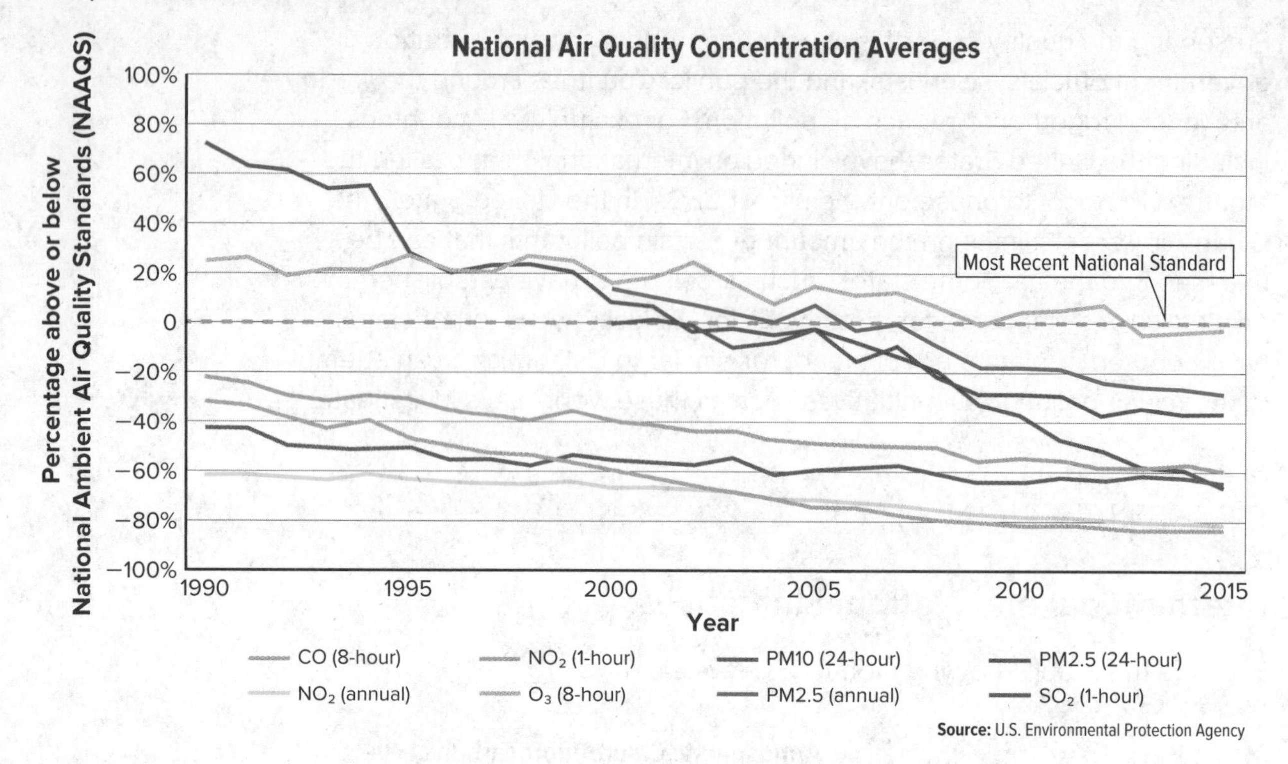

Source: U.S. Environmental Protection Agency

2. What trends or patterns do you notice in the data?

3. Would you consider the relationships between the data to be causal or correlational? For example, is the change in O_3 causing the change in PM10, or are the pollutants trending in parallel but one is not causing the other? Does correlation prove causation? Discuss your thoughts with your group.

4. After reviewing the data from both graphs, do you think treaties like the Montreal Protocol of 1987 and laws like the Clean Air Act of 1970 are an effective way to reduce air pollution? Use evidence from the investigation to support your claim.

Positive Actions While international, national, and state legislation helps monitor and minimize impacts on the atmosphere, people can also make a difference. Do you turn off the lights when you leave a room? If so, you are helping reduce your impact on the environment. You can help protect Earth's atmosphere in other ways as well, such as educating others about environmental issues, analyzing the choices you make as a consumer, and following some of the suggestions you will read about next.

INVESTIGATION

It's Your Turn

Study the figure below.

1. Can you identify ways in which this community is helping reduce air pollution?

2. What other methods or technologies would you suggest?

Making a Difference People have many options for reducing air pollution. You might be too young to own a house or a car, but you can perform simple activities such as planting trees to improve air quality, or turning off lights and electronic equipment when you are not using them to save energy. Walking, biking, or using public transportation also conserves energy. Recycling metal, paper, plastic, and glass reduces the amount of fuel required to manufacture these materials. Reducing energy use (both directly and indirectly) means that fewer pollutants are released into the air.

People also can invest in more energy-efficient technologies. Hybrid cars, for example, use both a battery and fossil fuels for power. Electric vehicles, such as the one shown below, use electric power alone.

Another way is to develop and use alternative sources of energy. Using renewable energy resources such as solar power, wind power, and geothermal energy to heat homes helps reduce air pollution. Recall that renewable resources are resources that can be replaced by natural processes in a relatively short amount of time.

What Gives? If solutions such as solar panels and wind turbines are engineered to help reduce air pollution caused by human activities, why aren't more individuals, towns, and countries relying solely on these types of technologies? In the following investigation, you will research and debate the advantages and disadvantages of Earth's energy resources.

ENGINEERING INVESTIGATION

Debate It

A town near you is looking to receive its energy from either nonrenewable or renewable resources. Your group will argue for or against using the different forms of energy.

1. Discuss with your teacher the procedures and rules for a debate, including rules for counterarguments.
2. **READING Connection** With your group, research the advantages and disadvantages of using the energy resource assigned to you by your teacher. Your teacher will also instruct whether you will argue for the advantages or disadvantages of your assigned energy resource. Record your research in your Science Notebook.
3. Identify any individual or societal needs, wants, and/or economic conditions that might influence the use or limitation of your energy resource. Include any other factors that might drive the use or limitation of technologies, such as the climate of your region, or the global distribution of your group's energy resource. Record your research in your Science Notebook.
4. **WRITING Connection** Prepare a presentation to support your stated opinions. Be sure to quote or paraphrase and properly cite your sources. Each team will present its arguments to the class.
5. After your team's presentation, listen carefully to the counterargument presentations given by the other teams. Take notes to record the different points of view for each energy resource. As you compare and critique the counterarguments, analyze whether they emphasize similar or different evidence or interpretations of facts.
6. After each argument has been presented, conduct a class survey to determine which team won the debate.

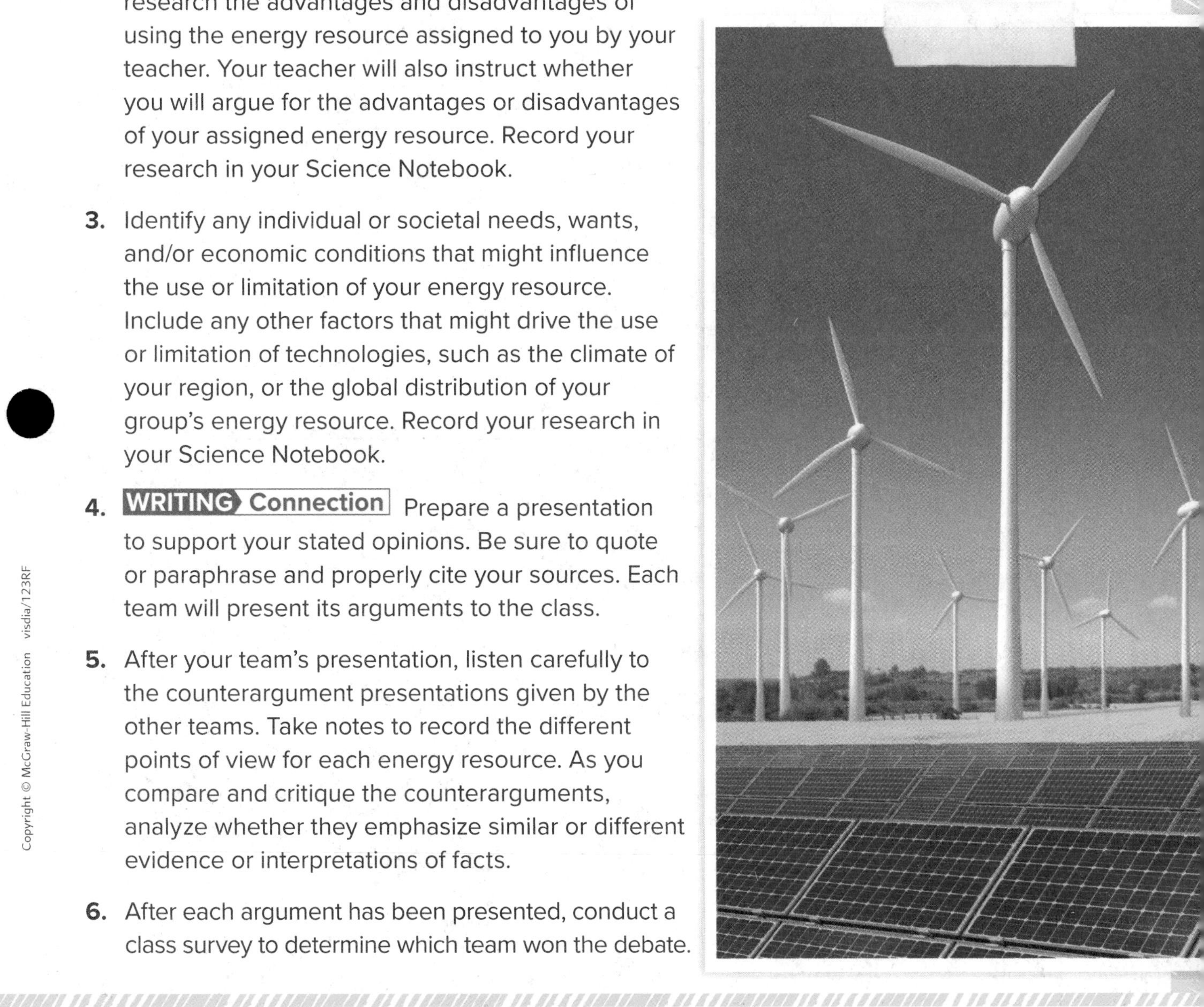

COLLECT EVIDENCE

How can air pollution be minimized? Record your evidence (B) in the chart at the beginning of the lesson.

LESSON 3
Review

Summarize It!

1. **Record** some of the negative and positive impacts that humans have on the atmosphere using the graphic organizer below.

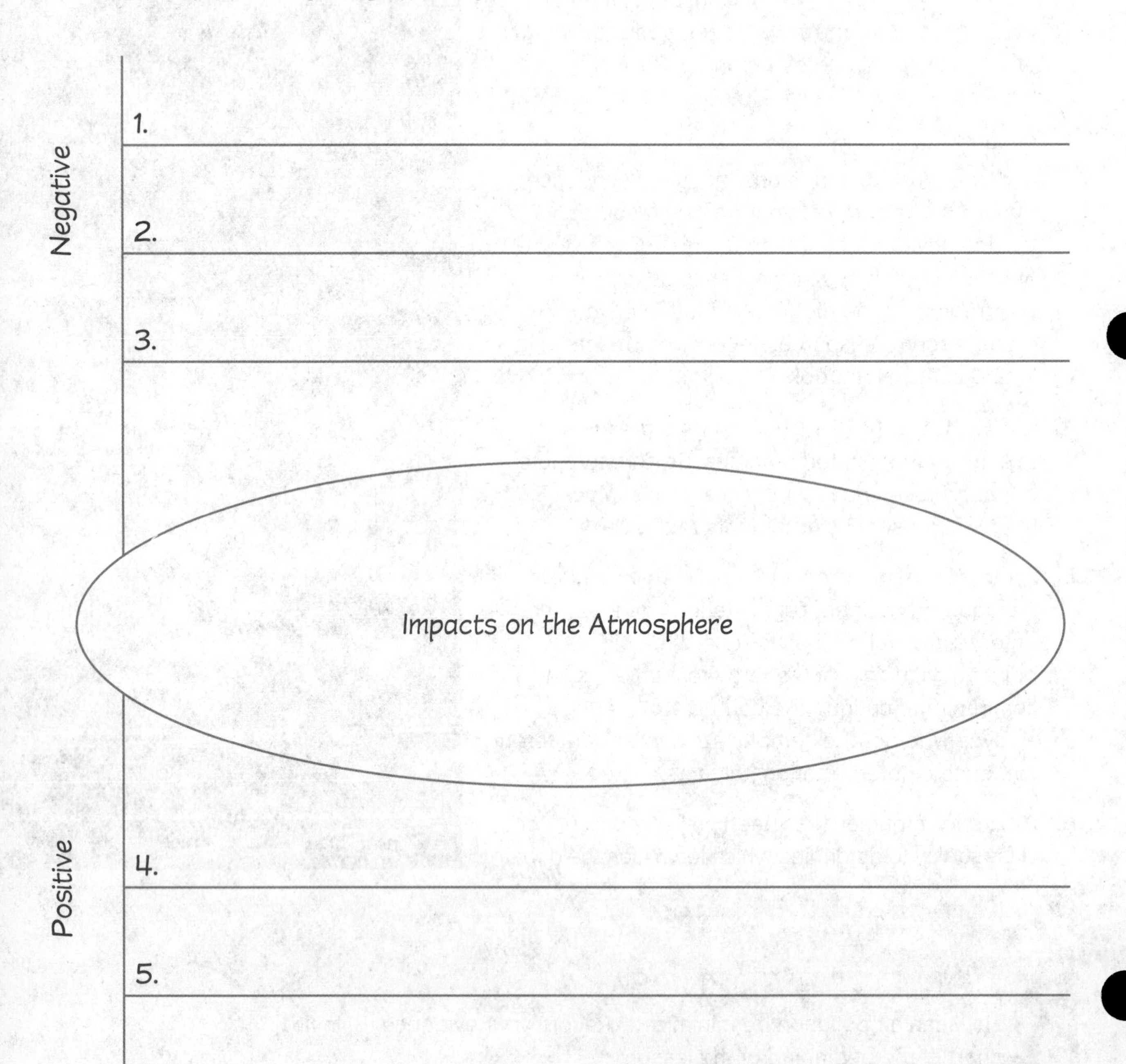

Three-Dimensional Thinking

Suppose your friend suffers from asthma. People with respiratory problems such as asthma are usually more sensitive to air pollution. *Sensitive* is a term used on the Air Quality Index (AQI). Use the AQI in the table below to answer the following question.

Air Quality Index			
Ozone Concentration (parts per million)	Air Quality Index Values	Air Quality Description	Preventative Actions
0.0 to 0.064	0 to 50	good	No preventative actions needed.
0.065 to 0.084	51 to 100	moderate	Highly sensitive people should limit prolonged outdoor activity.
0.085 to 0.104	101 to 150	unhealthy for sensitive groups	Sensitive people should limit prolonged outdoor activity.
0.105 to 0.124	151 to 200	unhealthy	All groups should limit prolonged outdoor activity.
0.125 to 0.404	201 to 300	very unhealthy	Sensitive people should avoid outdoor activity. All other groups should limit outdoor activity.

2. **HEALTH Connection** Today's AQI value is 130. What is the concentration of ozone in the air? Is today a good day for you and your friend to go to the park to play basketball for a few hours?

A Between 0.125 to 0.404 ppm; No. Sensitive people should avoid outdoor activity and all other groups should limit their activity outdoors.

B Between 0.0 to 0.064 ppm; Yes. The air quality is good and no preventative actions are needed.

C Between 0.105 to 0.124 ppm; No. All groups should limit prolonged outdoor activity.

D Between 0.085 to 0.104 ppm; No. Sensitive people should limit prolonged outdoor activity.

Real-World Connection

3. **Analyze** Why might smog be worse on a sunny weekday than on a sunny weekend?

4. **Describe** two ways you can personally help minimize human impact on the atmosphere.

Still have questions?
Go online to check your understanding about how human activities impact the atmosphere.

REVISIT SCIENCE PROBES

Do you still agree with the friend you chose at the beginning of the lesson? Return to the Science Probe at the beginning of the lesson. Explain why you agree or disagree with that friend now.

EXPLAIN THE PHENOMENON

Revisit your claim about why engineering solutions to minimize air pollution is important. Review the evidence you collected. Explain how your evidence supports your claim.

KEEP PLANNING

STEM Module Project Engineering Challenge

Now that you've learned about how humans impact the atmosphere, go back to your Module Project to continue planning your solution. Keep in mind that your goal is to design and evaluate a solution that will monitor and minimize an environmental impact of your choice.

PERFORMANCE EXPECTATION

PAGE KEELEY SCIENCE PROBES

LESSON 4 LAUNCH

Is Earth getting warmer?

Two students were talking about how cold it was during the past winter. They argued about climate change and whether they thought Earth was getting warmer.

Kirby: I think Earth is getting warmer. There is evidence that Earth's climate is changing, resulting in global warming.

Jan: I don't think Earth is getting warmer. It was freezing in Florida this year and there were big snowstorms in some places that normally get very little snow.

Circle the student you most agree with and explain why you agree. Describe your ideas about Earth's climate.

____

____

____

____

____

You will revisit your response to the Science Probe at the end of the lesson.

Impact on Climate

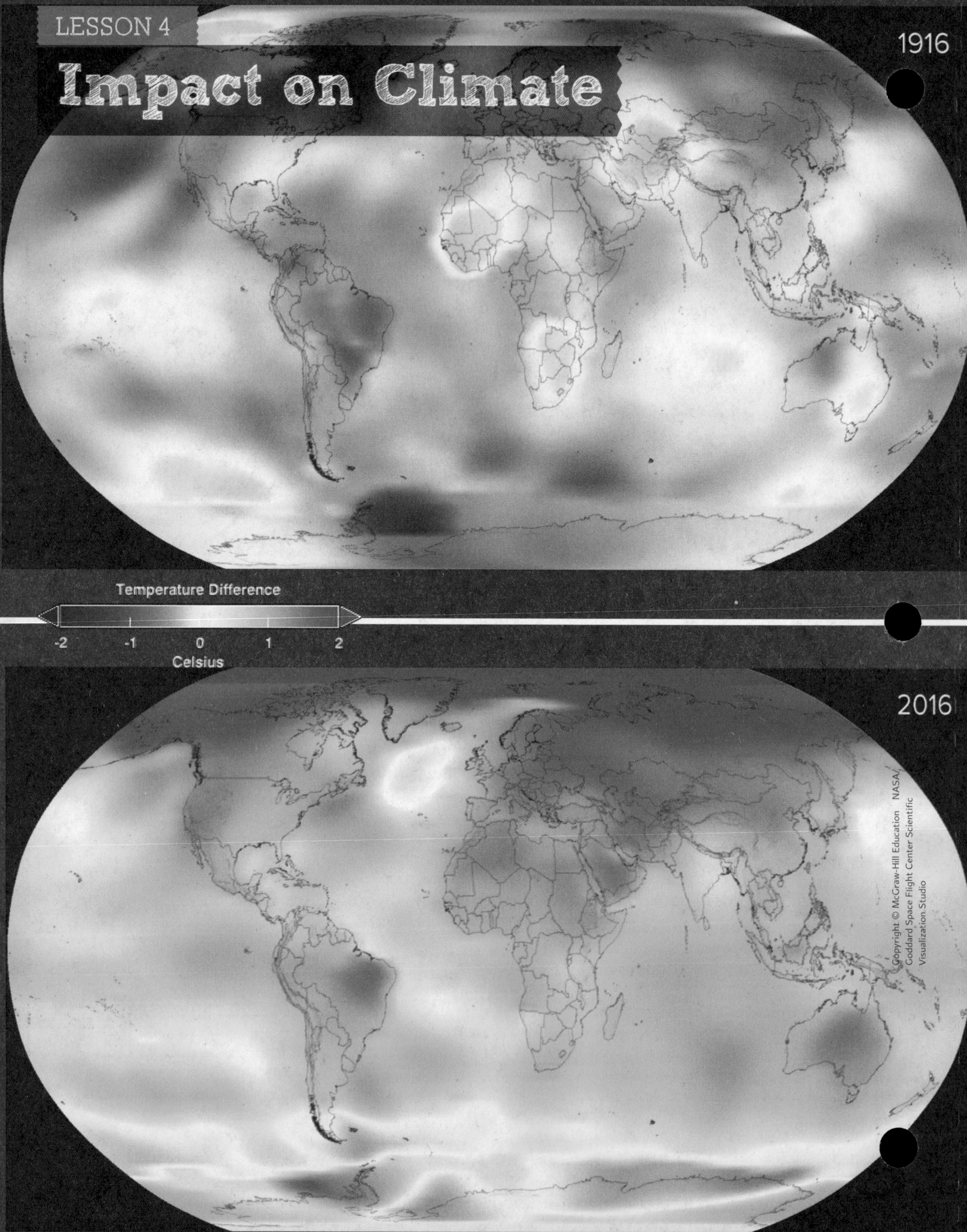

1916

Temperature Difference

-2 -1 0 1 2

Celsius

2016

ENCOUNTER THE PHENOMENON | What is happening to Earth's climate, and what is the impact?

GO ONLINE

Watch the video *Seeing Red* to see this phenomenon in action.

The color-coded maps in the video show how Earth's surface temperature has changed over time. Lower-than-normal temperatures are shown in blue. Higher-than-normal temperatures are shown in red. What did you notice as you progressed through the maps?

Brainstorm why this trend might be occurring and how it might impact Earth's systems. Record your thoughts below.

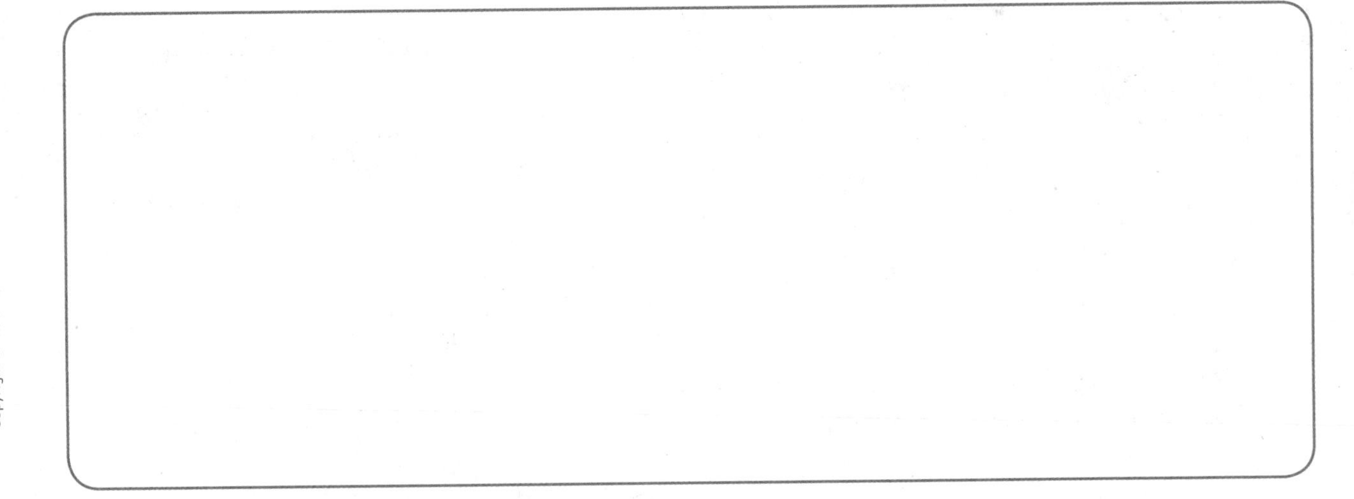

Now, with your partner determine three questions that you would ask a panel of climatologists to help clarify the phenomenon. Begin a "Question the Experts" list in your Science Notebook. You will add to this list of questions as you progress through the lesson.

EXPLAIN THE PHENOMENON

Are you starting to get some ideas about why global temperatures are changing and how this change influences Earth? Use your observations about the phenomenon to make a claim about the causes and effects of a warming planet.

CLAIM

The current rise in Earth's global temperatures is caused by... and affects Earth's systems by...

COLLECT EVIDENCE as you work through the lesson. Then return to these pages to record your evidence.

EVIDENCE

A. What evidence have you discovered to explain how greenhouse gases affect climate?

B. What evidence have you discovered to explain where excess greenhouse gases are coming from?

MORE EVIDENCE

C. What evidence have you discovered to explain how climate change affects Earth's systems?

D. What evidence have you discovered to explain how humans can minimize or adapt to the impacts of climate change?

When you are finished with the lesson, review your evidence. If necessary, based on the evidence, revise your claim.

REVISED CLAIM

The current rise in Earth's global temperatures is caused by... and affects Earth's systems by...

Finally, explain your reasoning for how and why your evidence supports your claim.

REASONING

The evidence I collected supports my claim because...

What is climate change?

In this module, you have learned about the ways in which humans impact Earth's land, water, and atmosphere. But did you know that human activities can also impact and change Earth's climate?

Climate is defined as long-term average weather conditions. Earth's climate has been changing since the planet formed 4.6 billion years ago, varying over time scales of decades to millions of years. **Climate change** can include global trends in warming, cooling, precipitation, wind directions, and other related measures. Although the phrases *climate change* and *global warming* are sometimes used interchangeably, **global warming** means a rise in Earth's average surface temperature. Global warming, therefore, is a part of climate change.

THREE-DIMENSIONAL THINKING

1. Is Earth's climate **stable?** Over what **time scales** can the **system change?**

2. **Explain** the difference between *climate change* and *global warming.*

Scientists agree that Earth's climate is changing. However, there are political disagreements about what should be done about it. What scientific evidence supports climate change?

Want more information?
Go online to read more about the factors that have caused the recent rise in global temperatures.

FOLDABLES®
Go to the Foldables® library to make a Foldable® that will help you take notes while reading this lesson.

What do temperature records show?

Thermometers provide a direct measurement of air temperature. Over the last 150 years, people have measured atmospheric temperatures. Let's investigate what this record shows.

INVESTIGATION

For the Record

Compare the temperature records of four international science institutions.

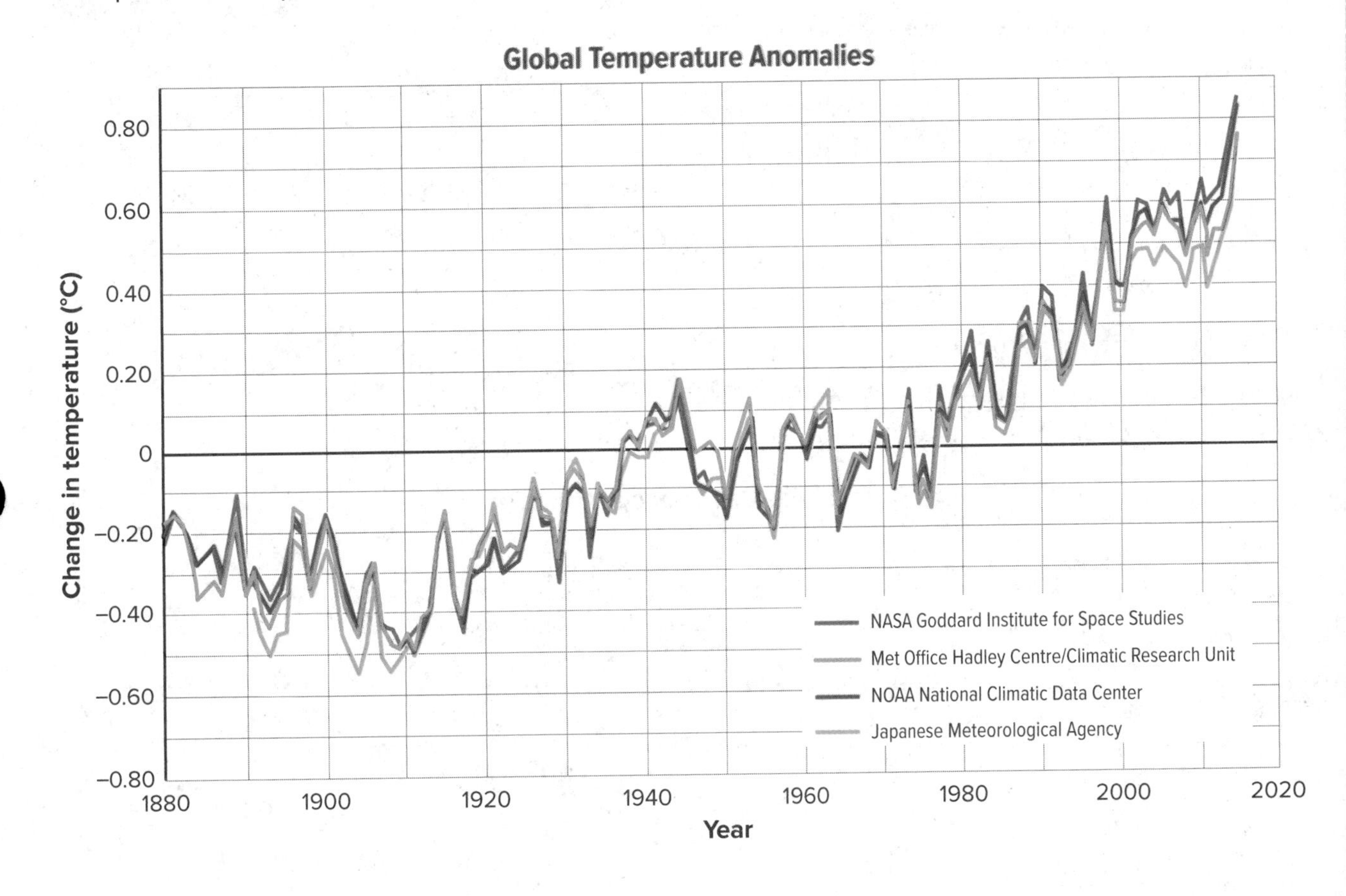

1. What patterns do you see in the data?

2. Do the data generally agree or disagree?

3. Ask your partner one or two questions about the data. Record your questions in your Science Notebook.

Scientific Consensus As you just investigated in *For the Record,* temperature records show that Earth is getting warmer. Climate scientists have been studying this change and the possible causes of it. Studies show that these changes are due to an increase in greenhouse gases in Earth's atmosphere.

What are greenhouse gases and how do they affect climate?

Gases in the atmosphere that absorb Earth's outgoing infrared radiation are called **greenhouse gases.** Carbon dioxide (CO_2) is an important greenhouse gas. Other greenhouse gases include methane (CH_4) and nitrous oxide (N_2O). What is happening to greenhouse gas concentrations in the atmosphere?

INVESTIGATION

Greenhouse Gases

Study the graph of atmospheric greenhouse gas levels determined from ice core data (dots) and from direct atmospheric measurements (lines).

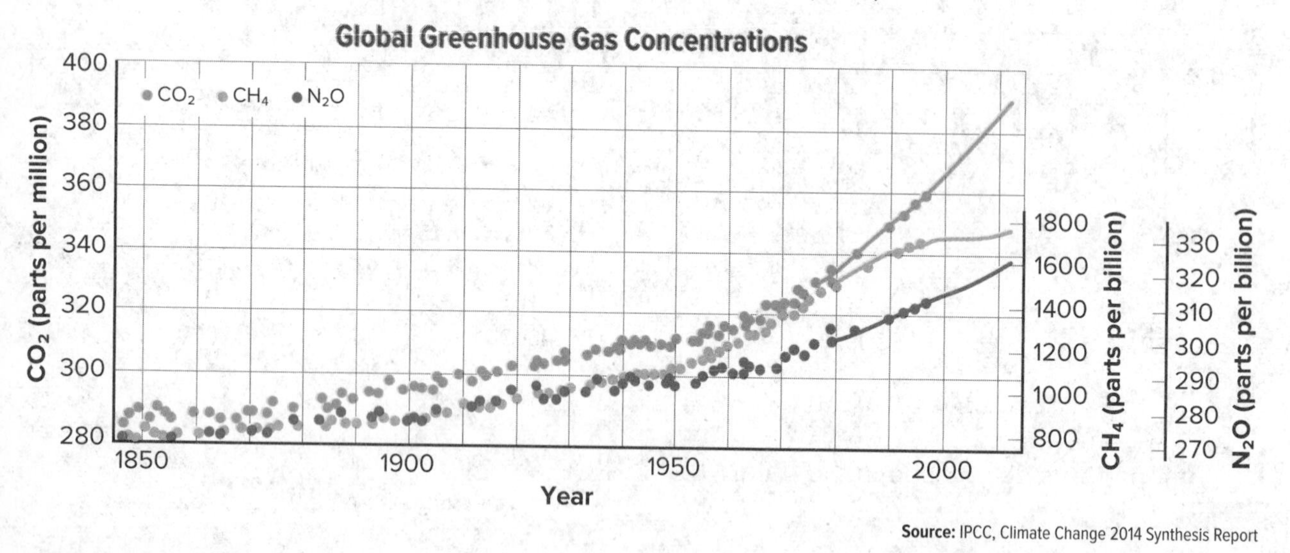

Source: IPCC, Climate Change 2014 Synthesis Report

1. What has happened to the levels of CO_2, CH_4, and N_2O in the atmosphere over the last century?

2. What is one question that you have about the data? Record your question in your Science Notebook.

GO ONLINE Now watch the animation *The Greenhouse Effect* to see how Earth's atmosphere acts a little like a greenhouse. Then answer the questions that follow.

3. What is the greenhouse effect?

4. Predict the effect of an increase in greenhouse gas concentrations in the atmosphere on Earth's average air temperatures.

Greenhouse Effect The **greenhouse effect** is the natural process that occurs when certain gases in the atmosphere absorb and reradiate thermal energy from the Sun. This thermal energy warms Earth's surface. Without the greenhouse effect, Earth would be too cold for life as it exists now.

However, higher levels of greenhouse gases create a greater greenhouse effect. When the amount of greenhouse gases increases, more thermal energy is trapped and Earth's surface temperature rises. Global warming occurs. Research shows that carbon dioxide is the most important of these gases in terms of climate change.

THREE-DIMENSIONAL THINKING

On the greenhouse below, **model** the greenhouse effect. Use yellow to illustrate incoming solar **energy.** Use orange to illustrate what happens to outgoing radiation.

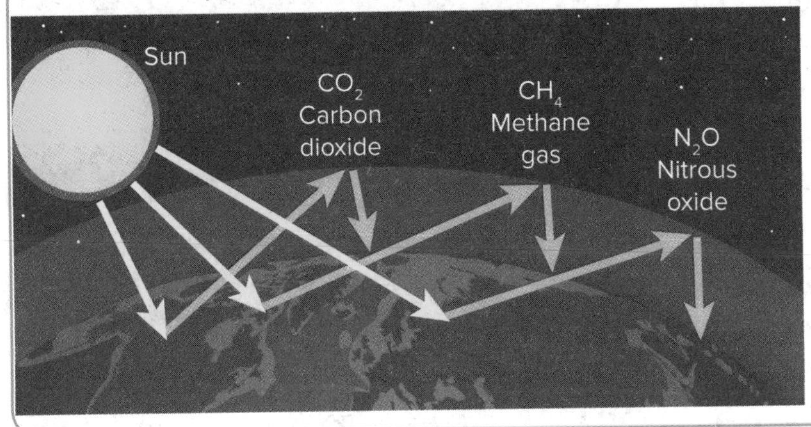

COLLECT EVIDENCE

How does an increase in greenhouse gases affect climate? Record your evidence (A) in the chart at the beginning of the lesson.

Where is all of the CO_2 coming from?

Volcanic eruptions, forest fires, and cellular respiration all naturally release carbon dioxide to the atmosphere. These sources alone, however, cannot explain why atmospheric carbon dioxide is currently at a level not seen for at least the last 800,000 years. Let's take a closer look.

Volcanic eruptions

Forest fires

Cellular respiration

INVESTIGATION

On the Rise

Study the graph showing data from atmospheric samples in ice cores and more recent direct measurements of CO_2.

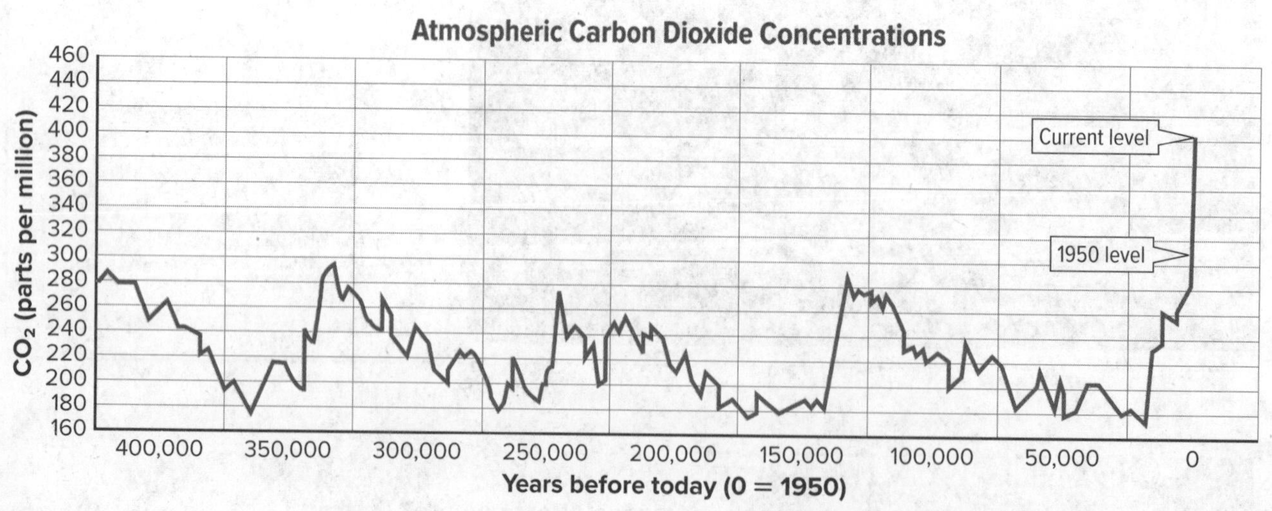

Source: Vostok ice core data/J.R. Petit et al.; NOAA Mauna Loa CO_2 record

1. How would you describe the data prior to 1950?

2. How would you describe the trend in atmospheric CO_2 from 1950 to now?

3. Formulate two questions that, if answered, would help clarify the data. Record these questions in your Science Notebook.

Now, compare the atmospheric greenhouse gas concentrations graph from the *Greenhouse Gases* investigation to the graph of human-produced CO_2 emissions below.

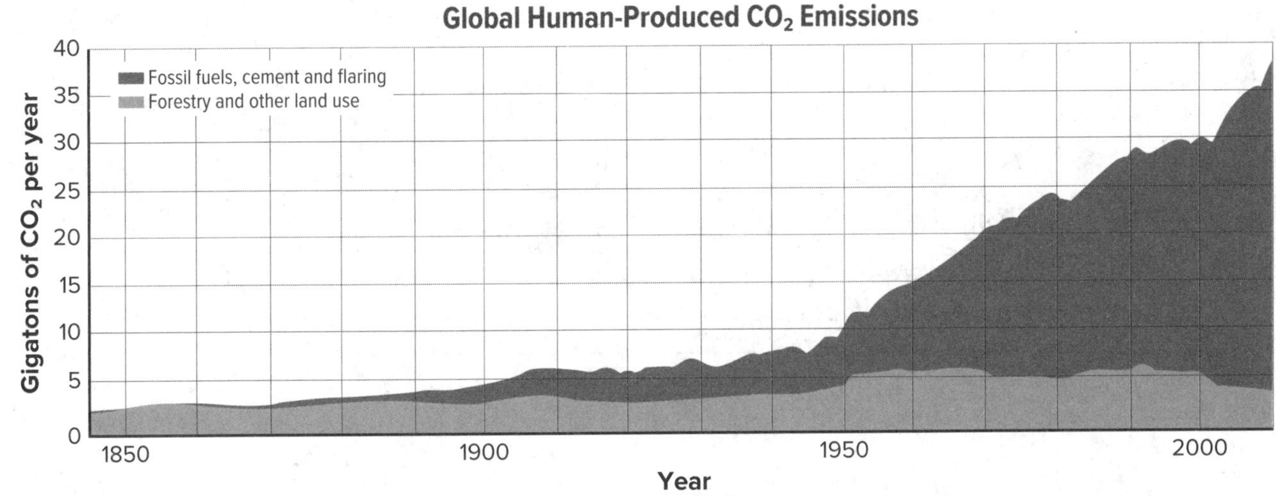

Source: IPCC, Climate Change 2014 Synthesis Report

4. What patterns do you notice between the two graphs?

5. What can you infer from this data?

6. A student argues that the best way to reduce human CO_2 emissions is to stop using cement for sidewalks and buildings. Using the graph above, pose a question to this student that challenges this argument. Record your question in your Science Notebook.

Finally, study the graph below showing the change in global temperature relative to 1986–2005 average temperatures. The colors indicate different datasets. Compare these to the previous datasets you have observed.

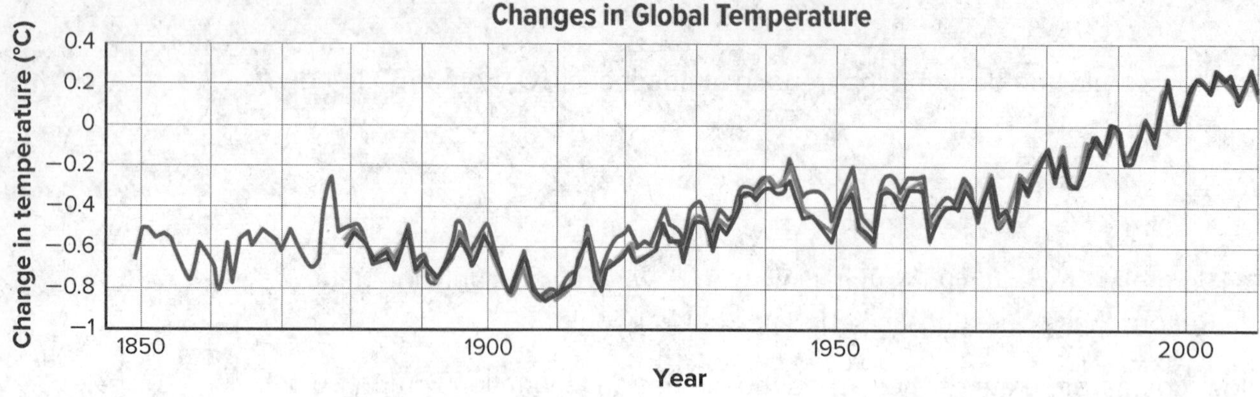

7. What patterns exist between the datasets?

8. **MATH Connection** What was the approximate temperature anomaly in 1910 versus in 2010? What is the approximate rate of change per decade between this timeframe?

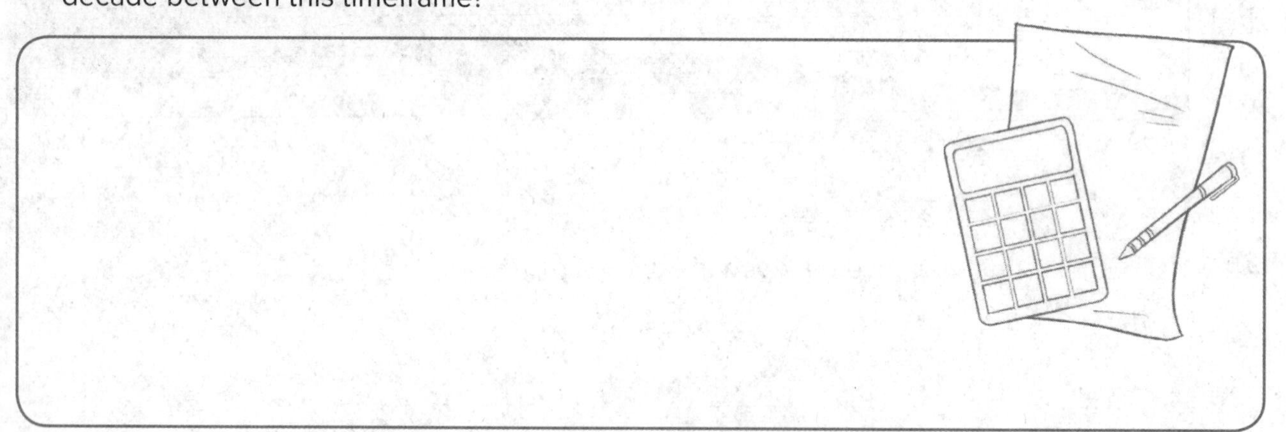

9. Would you consider the relationships as causal (an observed event or action appears to have caused a second event or action) or correlational (variables increase or decrease in parallel, however one might not have caused the other)? Explain.

Human Impact Are humans affecting the climate? According to 97 percent of actively publishing climate scientists—yes. In fact, according to the National Oceanic and Atmospheric Association (NOAA), over the last five decades natural factors would actually have led to a slight cooling of global temperatures.

In 2014, the Intergovernmental Panel on Climate Change (IPCC) concluded that the increase in CO_2 and other greenhouse gases and resulting climate change is extremely likely due to human activities. As you discovered in the *On the Rise* investigation, the main source of atmospheric carbon dioxide from humans is from the burning of fossil fuels. Deforestation and other land use is the second largest contributor of human-produced atmospheric CO_2.

SEP DCI CCC

THREE-DIMENSIONAL THINKING

Using what you have learned so far in the lesson, **explain** the conditions under which Earth's climate **system** can **change.**

GO ONLINE for an additional opportunity to explore!

ENGINEERING Connection Want to learn more about the greenhouse effect? Then perform the following activity.

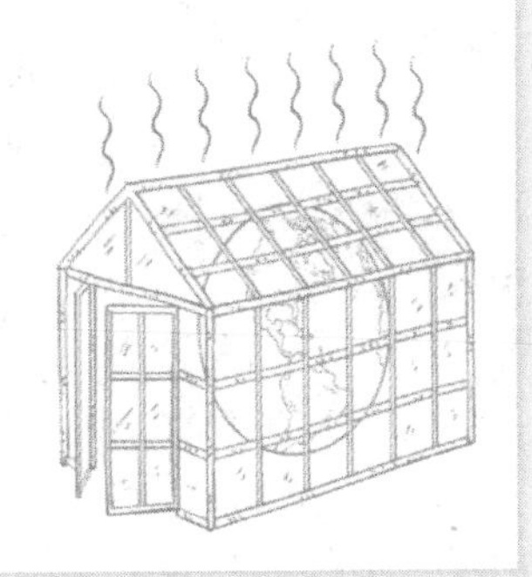

☐ **Develop** greenhouse models to test the greenhouse effect in the **Lab** *The greenhouse effect is a gas!*

COLLECT EVIDENCE

What is changing atmospheric levels of CO_2 and other greenhouse gases? Record your evidence (B) in the chart at the beginning of the lesson.

How is the recent warming trend affecting Earth's ice?

A changing climate can present serious problems for the environment and its organisms. Global warming has already had observable effects on Earth's natural systems. Let's take a closer look.

INVESTIGATION

Now You See It, Now You Don't

The Qori Kalis is a large glacier high in the Andes Mountains of Peru. Compare the glacier in 1978 to the glacier in 2011.

1978

2011

1. What has happened to the glacier over time?

2. Why do you think this occurred?

3. What consequences might arise because of phenomena such as this?

Melting Ice One of the most obvious signs of recent climate change is the retreat of glaciers around the world. Since 1980, glaciers with the longest, most reliable records decreased by an amount equivalent to 15.1 m (50 ft) of water. Glacier National Park, which had 150 glaciers in 1910, now has fewer than 30 glaciers left.

A glacier that spreads over land in all directions is called an ice sheet. Increased temperatures can also cause ice sheets to melt. For example, data from NASA's *GRACE* satellites show that the Greenland ice sheet has been losing an estimated 286 billion metric tons per year. Land ice sheets in both Greenland and Antarctica have seen an acceleration of ice mass loss since 2009.

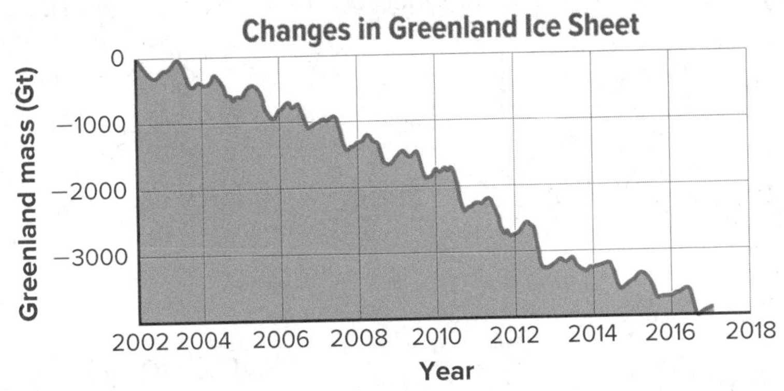

Source: National Aeronautics and Space Administration

Sea ice in the Arctic Ocean is also shrinking, as shown in the images below. The white surface of sea ice reflects up to 80 percent of incoming sunlight. With less sea ice present, the dark surface of the ocean absorbs much more of the Sun's energy, leading to further warming of the atmosphere. This leads to more melting of ice, which leads to further warming.

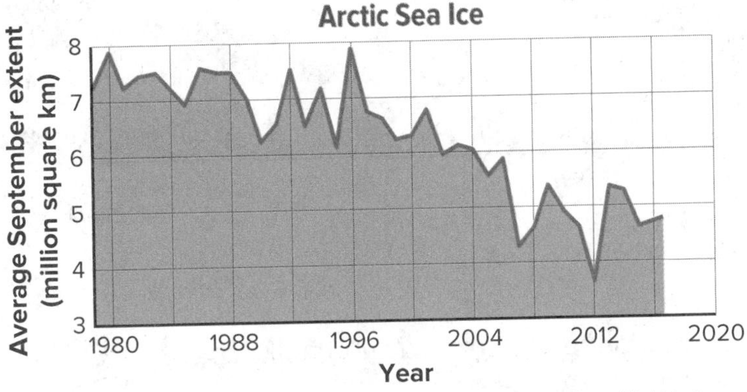

Source: National Aeronautics and Space Administration

1979

2016

How do you think warming temperatures and melting ice impacts Earth's oceans?

How is climate change affecting Earth's oceans?

The oceans absorb heat and carbon dioxide. As both global temperatures and atmospheric CO_2 are increasing, so are their effects on the oceans.

INVESTIGATION

Signs in the Sea

Examine the graph below.

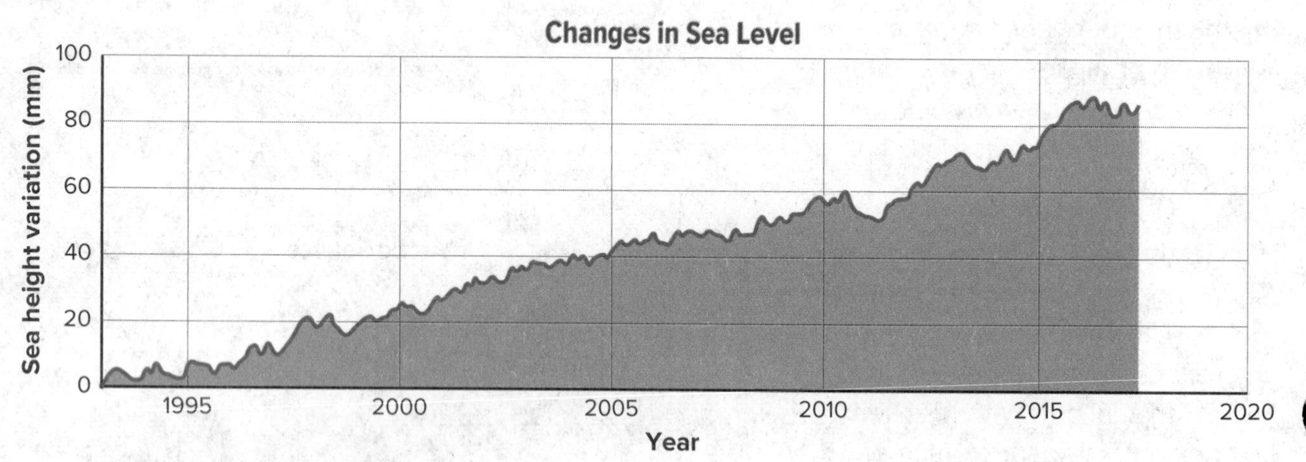

1. What pattern or trend do you notice?

2. **MATH Connection** The sea height variation was 15 mm in 1997 and 85 mm in 2017. Calculate the percent change in variation from 1997 to 2017.

Rising Seas As you just observed, the ocean's surface is rising. One of the major contributors to sea level rise is thermal expansion. When water warms, it expands and takes up more space. In addition, melting glaciers and polar ice sheets add more water to the oceans. Rising sea level erodes beaches and coastal wetlands and can flood low-lying areas. Coastal flooding is a serious concern for the 600 million people living in low-lying areas on Earth. Ecosystems can also be disrupted as coastal areas flood.

AMERICAN MUSEUM OF NATURAL HISTORY

CAREERS in SCIENCE

Oceans on the Rise—Again

With an eye to the future, a geologist examines past connections between higher sea levels and melting ice sheets.

▲ **This map shows coastal areas that would be flooded if the sea levels rise 6 m as scientists predict.**

Throughout its history Earth has gone through many long-term warming and cooling patterns that have lasted many thousands or millions of years. During cooling periods, much of the Earth was covered with ice sheets and glaciers. In warming periods the climate was mild, ice all but disappeared, and sea levels were higher. Scientists study these past climate patterns to understand the present and to predict what the future could bring.

Today about ten percent of the Earth's land mass is covered with ice sheets and glaciers. One vast ice sheet covers the island of Greenland in the Northern Hemisphere, while another covers the continent of Antarctica in the Southern Hemisphere. However, as global temperatures increase at an unprecedented rate due to human activity, the ice sheets are melting rapidly, causing sea level to rise. In addition, the ocean is warming, and thermal (heat) energy causes the water to expand. This expansion is another factor that contributes to sea level rise.

If the ice sheets continue to melt at this rate, scientists project sea levels will rise about 1.8–2.4 m (6–8 ft) by the year 2100. That's enough water to flood coastlines everywhere on Earth, threatening large cities and coastal habitats.

Coral lives and grows under water. Muhs shows where the sea level was in the past at this location in Florida. By measuring the height of the coral fossils, he estimates the ocean was once several meters higher than it is today. ▼

Daniel Muhs is a geologist with the United States Geological Survey. He studies rock layers for clues about past sea levels. He found one clue in the Florida Keys in a limestone wall that contains a fossilized coral reef. Today the wall sits several meters above sea level but it must have once been underwater for the reef to have grown there. Muhs determined that the reef grew there about 125,000 years ago during one of Earth's prior warming periods. Muhs estimates that sea levels back then were between 6 and 8 m (18 and 24 ft) higher than they are today. At that time the Earth was only about 2°C warmer than it is today. That indicates that it wouldn't take much of a temperature rise to dramatically raise sea levels. And since atmospheric warming is happening more quickly than in the past, the sea level rise could happen more quickly as well.

It's Your Turn

Research natural processes, such as changes in incoming solar radiation, that cause changes in Earth's climate over time. Could natural processes have caused the rise in global temperatures over the past century? Report your findings to the class.

LIFE SCIENCE Connection Besides contributing to sea level rise, warming oceans also have other consequences. Some marine organisms, such as coral, are very sensitive to temperature changes. A temperature increase as small as 1°C can cause corals to die. **Coral bleaching** is the loss of color in corals that occurs when stressed corals expel the algae that live in them. These algae provide the coral organisms with a source of matter and energy. Coral bleaching can kill corals, which in turn destroys important habitats for fish and many other organisms.

The skeletons of corals and the shells of many other organisms are also affected by another issue with climate change: carbon dioxide.

PHYSICAL SCIENCE Connection In addition to warming and rising seas, the oceans are becoming more acidic. Why? O_2 and CO_2 gases move freely between the atmosphere and seawater. As the amount of CO_2 increases in the atmosphere, the amount of CO_2 dissolved in seawater also increases. When CO_2 mixes with seawater, a weak acid called carbonic acid forms. Carbonic acid lowers the pH of the water, making it slightly acidic.

How does carbonic acid affect marine organisms? Let's dive in.

LAB An "Eggcellent" Question

Safety 

Materials

brown eggshell
plastic cup
white vinegar
forceps

Procedure

1. Read and complete a lab safety form.
2. Examine a piece of brown eggshell and describe its properties in the table on the right.
3. Place the eggshell in a plastic cup.
4. Half fill the cup with white vinegar.
5. After 15 minutes, use forceps to remove the eggshell. Describe its properties in the table.
6. Follow your teacher's instructions for proper cleanup.

Calcium-Containing Shells		
Property	**Description Before Treatment**	**Description After Treatment**
Hardness		
Thickness		
Appearance		

Analyze and Conclude

7. How might long-term effects of increased CO_2 in seawater affect calcium-containing shells and skeletons of marine organisms? Use evidence from the lab to support your claim. Write your response in your Science Notebook.

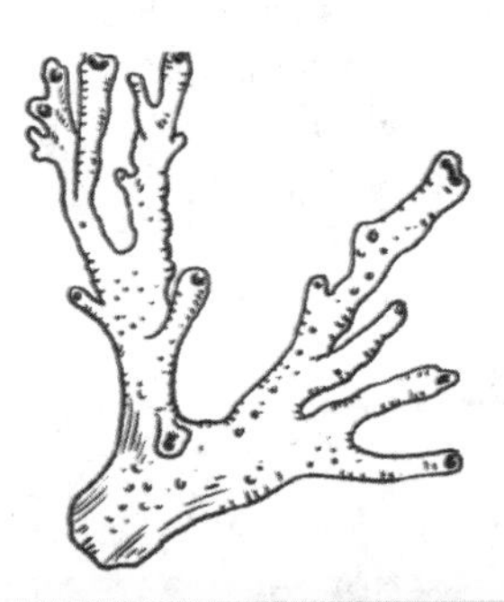

Ocean pH Absorbed CO_2 is acidifying the oceans. Many marine organisms build shells and skeletons from calcium absorbed from seawater. As seawater becomes more acidic, it is harder for these organisms to absorb calcium. As you just explored in the Lab *An "Eggcellent" Question,* increased acidity can cause shells and skeletons to weaken or dissolve. This could potentially lead to a loss of biodiversity and marine resources.

What does a warming climate mean for ecosystems on land?

LIFE SCIENCE Connection The geographic distribution of organisms can be significantly altered by climate change. As climates warm, organisms that were formerly restricted to warmer regions will become more common toward the poles.

Read a Scientific Text

CLOSE READING

Inspect

Read the the passage *Shifting Ecosystems.*

Find Evidence

Reread the third paragraph. Highlight the positive and negative impacts discussed.

Make Connections

Communicate With your partner, discuss how a warming climate can affect the seasonal migrations or life cycles of organisms. Have you noticed any changes in your region?

Shifting Ecosystems

Biologists report that many animals are breeding earlier or extending their ranges into new territory as the climate changes. In Europe and North America, for example, 57 butterfly species have either died out at the southern end of their range, or extended their northern limits, or both.

Plants also are moving into new territories. Given enough time and a route for migrations, many species may adapt to new conditions, but we now are forcing many of them to move much faster than they moved at the end of the last ice age. Trees such as oaks typically migrate at a rate of a few hundred meters per year, about one-tenth of the rate of temperature over land.

Insect pests and diseases have also expanded their range as hard winters have retreated northward. Mosquitoes, for example, have spread as warmer temperatures heat up environments that were previously too cold. While this increase in habitat is beneficial for organisms that use these insects as a food source, mosquitoes are carriers of infectious diseases.

COLLECT EVIDENCE

How is climate change affecting Earth's systems? Record your evidence (C) in the chart at the beginning of the lesson.

What can be done?

In this lesson, you explored the science behind the causes and effects of rising global temperatures. Understanding the climate system is essential to understand the ways in which we can minimize the impact of human activities. We can choose to pursue mitigation strategies that slow the pace of climate change or adaptation strategies that help us to adjust to its impact.

THREE-DIMENSIONAL THINKING

1. Using your understanding of the **causes** of climate **change**, brainstorm three methods that might help **stabilize** Earth's climate. How would these activities or technologies help mitigate global warming?

2. Using your understanding of the **effects** of climate change, brainstorm three ways to adapt to these effects and reduce human vulnerability.

3. With your partner, formulate two **questions** that you have about mitigating or adapting to climate change. Record your questions in your Science Notebook.

Addressing Climate Change Reducing the level of climate change might seem challenging. However, solutions are possible if we choose to act as individuals and as a society. In the previous lessons, you learned about ways to reduce human impact on Earth's land, water, and air. Many of these same activities and technologies can help reduce human impact on climate as well.

ENGINEERING INVESTIGATION

Envisioning Solutions

1. **READING Connection** Below are some of the ways humans might reduce the level or rate of climate change. With your team, choose one bullet point to research and report on. How costly is your method? Is your solution already in use? Is it working? Use the Internet and library to gather relevant sources to answer these and any other questions you might have about your method. Don't forget to assess the credibility of each source.

   - International agreements such as the Kyoto Protocol and the Paris Agreement
   - Alternative energy sources like wind, solar, hydroelectric, geothermal, and nuclear power
   - Conservation (cutting back on energy use) and efficiency (updating current systems that use energy)
   - Carbon capture and storage (also called carbon sequestration)
   - Increasing natural carbon sinks through forest management and conservation tillage

2. How likely do you think it is that society will accept or practice your method? How does understanding human behavior play into your thinking? Record your thoughts in your Science Notebook.

3. **WRITING Connection** What additional questions arose after researching your method? Record these in your Science Notebook.

4. Present your findings to the class.

COLLECT EVIDENCE

How can we minimize human impact on climate change? Record your evidence (D) in the chart at the beginning of the lesson.

Adapting to Climate Change Although strategies and activities to address climate change today are important, it could be 100 years or more before the effects of today's actions are realized. Meanwhile, societies must adjust to increasingly warmer climates. Some societies already are doing just that. In France, farmers are planting crops that can tolerate warmer temperatures. In Austria, ski resorts are planning hiking trails and golf courses for a future without snow. In Copenhagen, Denmark, new subway systems will be built with raised structures that allow for the rise in sea level that is expected from global warming. The table below shows a few more strategies.

Adjustment Strategies	
Agriculture	• Change planting dates • Change livestock and/or crops
Health	• Develop heat-health plans • Increase climate-related disease programs
Infrastructure	• Relocate at-risk communities in low-lying regions • Expand climate/weather monitoring systems

The question, however, remains: can entire societies really adjust? The answer might depend on how warm the climate gets and whether societies can change as quickly as the climate does.

INVESTIGATION

Stewards of the Planet

GO ONLINE to watch the video *Sustainable Earth*.

Discuss with your partner what you think the main "take away" is of the video. Record your thoughts below.

LESSON 4

Review

Summarize It!

1. **Explain** Return to your "Question the Experts" list in your Science Notebook. Choose your top three questions that, when answered, help clarify evidence of recent climate change. Think about how you can phrase the questions so that you get a comprehensive answer, rather than a simple "yes" or "no" response. Trade your questions with a partner. In the space below, record and answer your partner's questions using evidence from the lesson or additional research if needed.

Three-Dimensional Thinking

Analyze the graph below. Then answer the questions that follow.

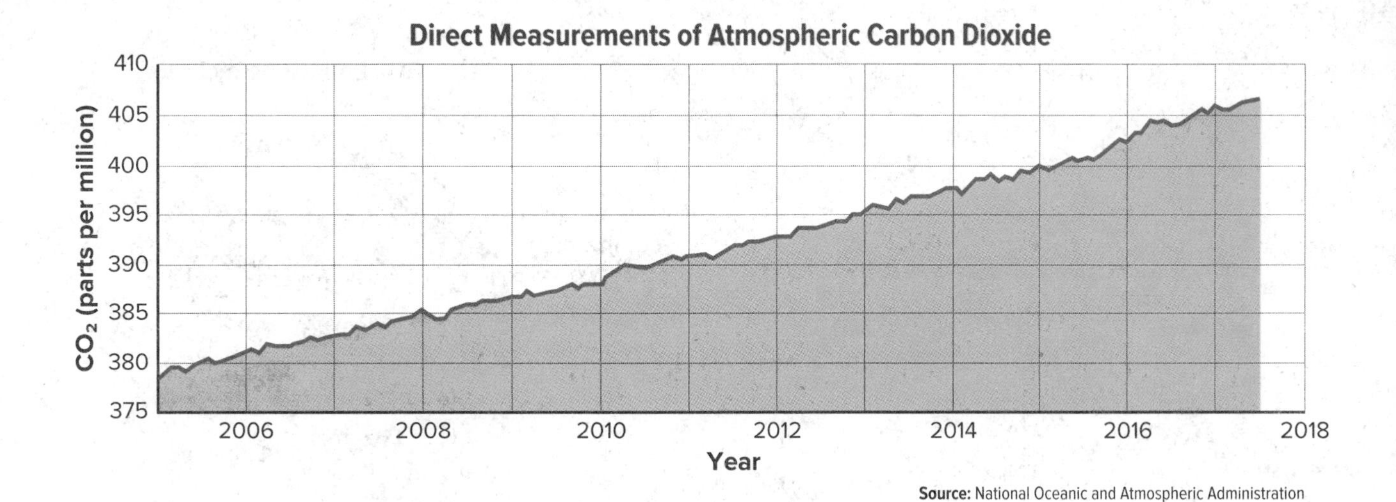

2. **MATH Connection** What is the approximate rate of change in atmospheric carbon dioxide from 2007 to 2017?

A 2.3 ppm per year

B 1.5 ppm per year

C 0.7 ppm per year

D 4.6 ppm per year

3. Which of the below conclusions can you draw from the data?

A Carbon dioxide is an important greenhouse gas that leads to global warming.

B When the amount of greenhouse gases increases, Earth's surface temperature rises.

C Direct measurements of carbon dioxide show a yearly increase since at least 2005.

D Human and natural activities have caused the rise in atmospheric carbon dioxide.

Real-World Connection

4. **Analyze** the graph. What are the top two sources of atmospheric CO_2 in the United States? Describe two ways that you could help to reduce these emissions and explain why activities like these are important.

U.S. CO_2 Emissions by Source

Source: U.S. Environmental Protection Agency

Still have questions?
Go online to check your understanding about how human activities impact global temperatures.

REVISIT

Do you still agree with the student you chose at the beginning of the lesson? Return to the Science Probe at the beginning of the lesson. Explain why you agree or disagree with that student now.

EXPLAIN THE PHENOMENON

Revisit your claim about the causes and effects of climate change. Review the evidence you collected. Explain how your evidence supports your claim.

PLAN AND PRESENT

STEM Module Project Engineering Challenge

Now that you've learned about how humans impact the climate, go back to your Module Project to continue planning your solution, design and evaluate your solution, and present it to the town committee. Your goal is to design a method for monitoring and minimizing a human impact on the environment.

PERFORMANCE EXPECTATION

Who's moving in next door?

A developer would like to purchase a large area of farmland near a small town. She wants to build a shopping mall on the land. Community members are worried about the environmental impacts that the development may have on their area. They want to know how it will affect land, water, and air resources. A committee has been formed to research these impacts before any decisions are made.

You work for an environmental consulting firm. Your team's task is to design plans for monitoring and minimizing environmental impacts and submit those plans to the town committee.

Planning After Lesson 1

In your own words, clearly describe your Engineering Challenge.

Define the boundaries of your design in terms of the environmental impacts on land. How will the shopping mall impact land resources? Consider both living and nonliving things.

Planning After Lesson 1, continued

What design solutions have been developed to monitor and minimize the environmental impact of similar developments on the land?

Planning After Lesson 2

Define the boundaries of your design in terms of the environmental impacts on water. How will the shopping mall impact water quality?

What design solutions have been developed to monitor and minimize the environmental impact of similar developments on water quality?

Planning After Lesson 3

Define the boundaries of your design in terms of the environmental impacts on air. How will the shopping mall impact air quality?

What design solutions have been developed to monitor and minimize the environmental impact of similar developments on air quality?

STEM Module Project
Engineering Challenge

Planning After Lesson 4

Define the boundaries of your design in terms of the environmental impacts on climate change. Will the shopping mall contribute to climate change? Explain.

What design solutions have been developed to monitor and minimize the environmental impact of similar developments on climate change?

Design Your Plan

Review the design challenge and the system boundaries you identified after each lesson. Use this information to design your plan. Be sure to include criteria and constraints.

Sketch Your Plan

In the space below, make a sketch of the shopping mall with your chosen solutions incorporated into the design of the development.

Evaluate Your Plan

Next, use the chart below to evaluate your plan.

	Yes	No	Proposed Revision to Plan
Did you distinguish between causal and correlational relationships in your plan?			
Does the plan meet your criteria?			
Does the plan meet your constraints?			
Did you identify limitations of the use of technologies in your plan?			

Submit Your Plan

After you have evaluated your plan, submit it to your classmates, who will play the role of the town committee. Answer any questions they may have about your plan. Then answer the questions below.

Was any one solution able to address all the potential environmental impacts of the shopping mall? If not, how did you solve this problem?

All human activity has both short- and long-term consequences, positive as well as negative, for the health of people and the environment. With this in mind, describe the consequences of building the shopping mall.

Based on what you've learned in this activity, what do you think drives the use of technology in any particular place?

Congratulations! You've completed the Engineering Challenge requirements!

Module Wrap-Up

Using the concepts you've learned in the module, explain how human activities impact Earth's land, water, atmosphere, and climate.

OPEN INQUIRY

If you had to ask one question about what you studied, what would it be?

Plan and conduct an investigation to answer this question.

Earth and Human Activity

ENCOUNTER THE PHENOMENON

What can patterns of light at night tell us about the human population and resource use?

Earth at Night

GO ONLINE
Watch the video *Earth at Night* to see this phenomenon in action.

Communicate The satellite image shows Earth at night. You can see where the large cities are located. What do you think the dark areas represent? When you turn on the lights at night, where does the energy to power the lights come from? How might this daily activity impact the environment? Record your thoughts in the space below. With a partner, discuss his or her thoughts. Then record what you both would like to share with the class below.

STEM Module Project Launch
Science Challenge

7.6 Billion and Counting

Scientists estimate that more than 7.6 billion people currently live on Earth. It is estimated that by 2050, the human population will reach 9.8 billion. How do the human population and human activities impact Earth?

You are on a team of scientists who are trying to answer this question. You will gather evidence to support an argument about how increases in human population and resource use affect Earth's systems. You will present your argument in a panel discussion with your team.

Start Thinking About It

In the image above you see a crowded city street, showing just one tiny portion of the 7.6 billion people on Earth. Think about how many resources you use in one day. How would Earth's systems be affected if one person's daily resource use were multiplied by 7.6 billion people? Discuss your thoughts with your group.

STEM Module Project

Planning and Completing the Science Challenge
How will you meet this goal? The concepts you will learn throughout this module will help you plan and complete the Science Challenge. Just follow the prompts at the end of each lesson!

LESSON 1 LAUNCH

Population Fluctuation

Three friends were arguing about how the human population has changed over time.

Casey: The human population had been fairly constant throughout Earth's history.

Allayna: Population size is increasing at a significant rate due to advances in technology, medicine, and sanitation.

Karl: Population size has been decreasing over time because people are having fewer children.

Which person do you agree with the most? Explain your reasoning.

You will revisit your response to the Science Probe at the end of the lesson.

LESSON 1

Human Population Growth

Copyright © McGraw-Hill Education Danita Delimont/Gallo Images/Getty Images

ENCOUNTER THE PHENOMENON

Why are the numbers of cars on the road increasing, and what's the impact?

GO ONLINE

Check out *On the Road* to see this phenomenon in action.

Using the information in the infographic online, create a graphic organizer that shows the relationships between human population growth and the consumption of natural resources. Display your graphic organizer in the space below.

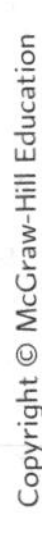

EXPLAIN THE PHENOMENON

Are you starting to get some ideas about how increases in the size of the human population causes increases in the consumption of natural resources? Use your observations about the phenomenon to make a claim about how the rate of consumption of natural resources is impacted by human population growth.

CLAIM

The relationship between human population growth and resource consumption is...

COLLECT EVIDENCE as you work through the lesson. Then return to these pages to record your evidence.

EVIDENCE

A. What evidence have you discovered to explain how the human population has changed over time?

MORE EVIDENCE

B. What evidence have you discovered to explain the impact of population growth on Earth's resources?

When you are finished with the lesson, review your evidence. If necessary, based on the evidence, revise your claim.

REVISED CLAIM

The relationship between human population growth and resource consumption is...

Finally, explain your reasoning for how and why your evidence supports your claim.

REASONING

The evidence I collected supports my claim because...

How fast is the population growing?

You are part of a population of humans. The other species in your area, such as birds or trees, each make up a separate population. A **population** is all the members of a species living in a given area. How has human population changed over time? Let's investigate.

INVESTIGATION

Here We Grow Again

Analyze the information presented in the graph.

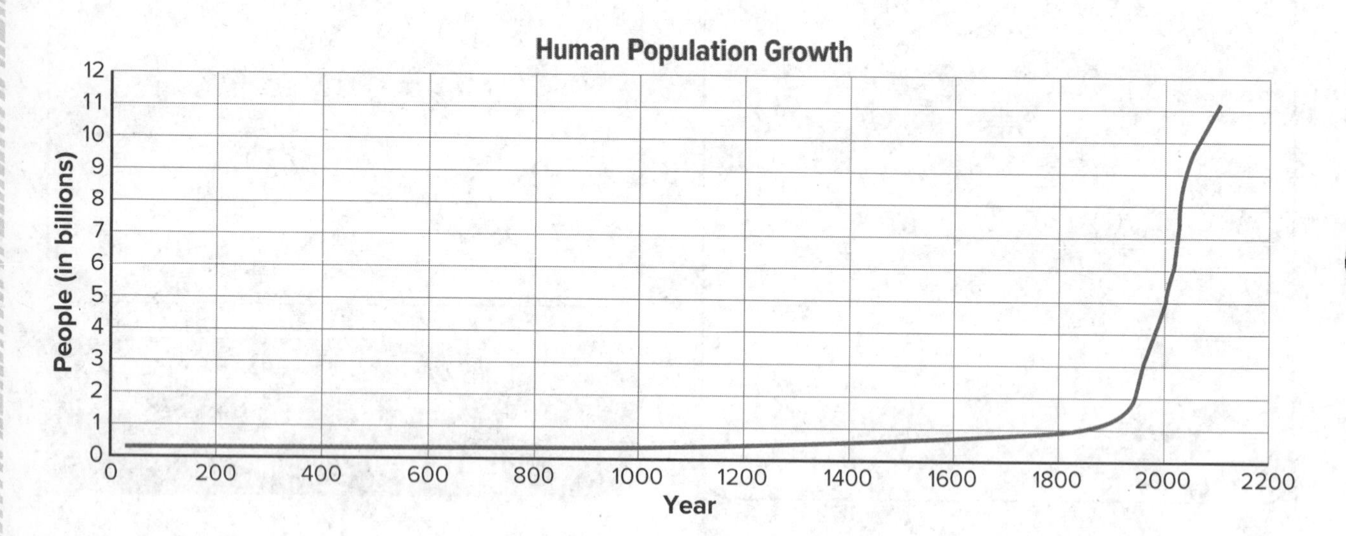

1. How does the rate of human population growth from years 200 to 1800 compare to the rate of growth from 1800 to 2000?

2. **WRITING Connection** Research the Industrial Revolution in the 1800s. Then, evaluate the following claim: The Industrial Revolution led to a dramatic population explosion. Construct an argument to support or refute the claim using evidence from your research. Record your response in your Science Notebook.

Human Population When the first American towns were settled, most had low populations. Today, some of those towns are large cities, crowded with people. In a similar way, Earth was once home to relatively few humans. Scientists estimate that there were about 300 million humans on Earth a thousand years ago. Improved health care, clean water, and other technological advancements mean that more people are living longer and reproducing.

As you discovered in the *Here We Grow Again* investigation, the greatest increase in human population occurred during the last few centuries. Today there are more than 7 billion humans on Earth. By 2050 there could be over 9 billion.

COLLECT EVIDENCE

How has human population changed over time? Record your evidence (A) in the chart at the beginning of the lesson.

GO ONLINE for additional opportunities to explore!

Want to learn more about population growth? Then perform one of the following activities.

- [ ] **Explore** how the population is projected to change in the **Investigation** *The Human Population.*

OR

- [ ] Analyze trends in human population data in the **Investigation** *Graphing Population Growth.*

Want more information?
Go online to read more about how resource availability is affected by a growing population.

FOLDABLES®
Go to the Foldables® library to make a Foldable® that will help you take notes while reading this lesson.

Is human population growth equal around the world?

In approximately one hour, 15,000 babies are born worldwide. Technological advances have resulted in a rapid growth in human population. Is human population growth equal in all countries? Let's investigate.

INVESTIGATION

Comparing Ages

Observe the graph showing the percent of youth and elderly in four countries.

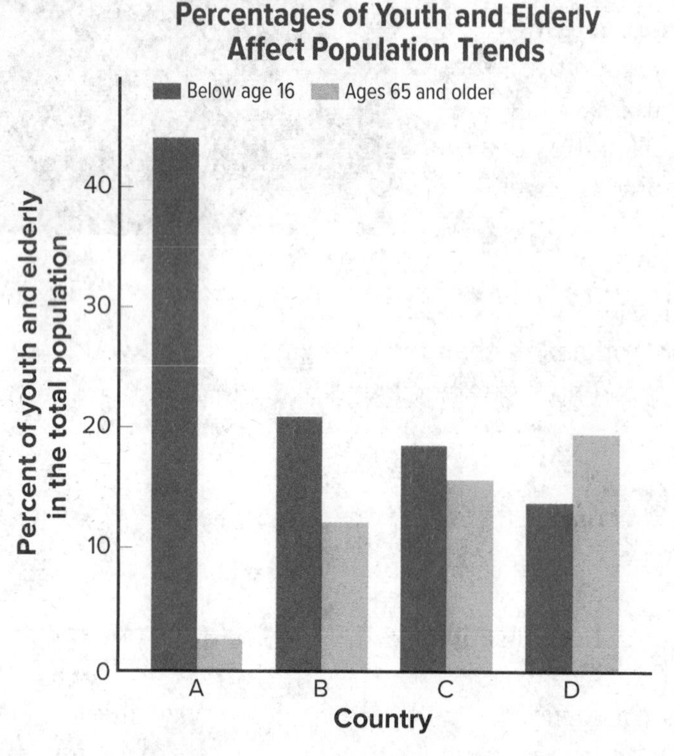

1. Use the data to predict how the population of each country will change.

2. Brainstorm a list of factors, events, or conditions that affect the growth of human populations in these countries. Predict the effect of each factor on the population growth rate.

Population Structure Populations have a tendency to increase in size. However, many factors influence the rate at which a population can grow. At the simplest level, the rate of increase is determined by subtracting the number of individuals leaving a population from the number entering the population. Individuals leave a population either by death or emigration, which is when people leave an area to settle somewhere else; they enter the population by birth or immigration, which is when people enter an area to live there. Another important characteristic of any population is its age structure. Let's investigate what an age structure diagram can tell us about population growth.

INVESTIGATION

Increasing and Decreasing Populations

The diagram below shows the relative numbers of individuals in pre-reproductive, reproductive, and post-reproductive years for three countries: Kenya, the United States, and Germany.

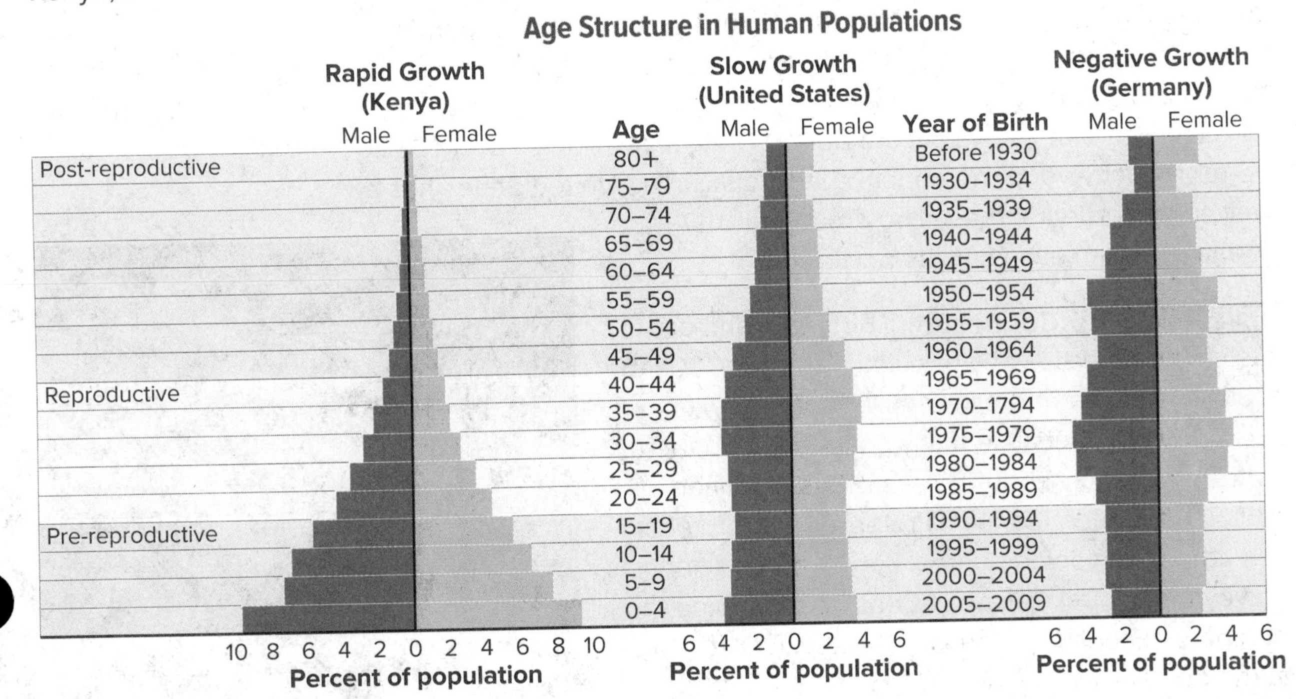

1. How would you describe Kenya's growth?

2. Why does a chart like Germany's show negative growth?

3. What can you determine about population growth rates from observing the top halves of the three charts?

Age Structure A population's age structure is the number of males and females in each of three age groups: pre-reproductive state, reproductive state, and post-reproductive state. The age structure diagrams you just analyzed are typical of many countries in the world. A nongrowing population looks like a rectangle. A slow-growing population looks like a rectangle with a bulge in the middle. A rapidly growing population looks like a triangle with its base at the bottom. Kenya has a large portion of pre-reproductive and reproductive individuals. The United States has a smaller proportion of these two groups, and Germany has an even smaller portion.

LIFE SCIENCE Connection Birth rate and death rate change the size of a population. In the 1700s the death rate of sea otters in central California was extremely high because many people hunted them. By the 1930s only about 50 sea otters remained. Today, the Marine Mammal Protection Act protects sea otters from being hunted. Every spring, scientists survey the central California Coast to determine the numbers of adult and young sea otters (called pups) in the population.

THREE-DIMENSIONAL THINKING

This graph shows the world population growth and projected trends in both developed and developing countries.

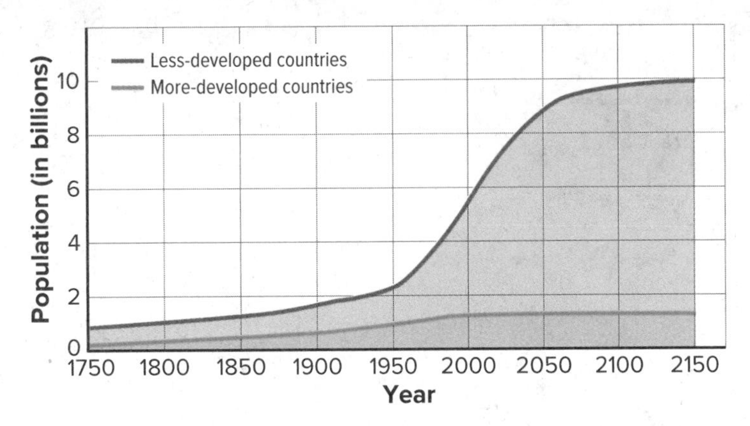

Analyze the data to determine the current population structure of developed and developing countries. Support your reasoning with evidence from the graph.

How does human population growth affect natural resources?

Every human being needs certain things, such as food, clean water, and shelter, to survive. People also need clothes, transportation, and other items. All the items used by people come from resources found on Earth. A **natural resource** is a part of the environment that supplies material useful or necessary for the survival of living things. Let's investigate how natural resources are affected by a growing population.

LAB Bean There, Done That

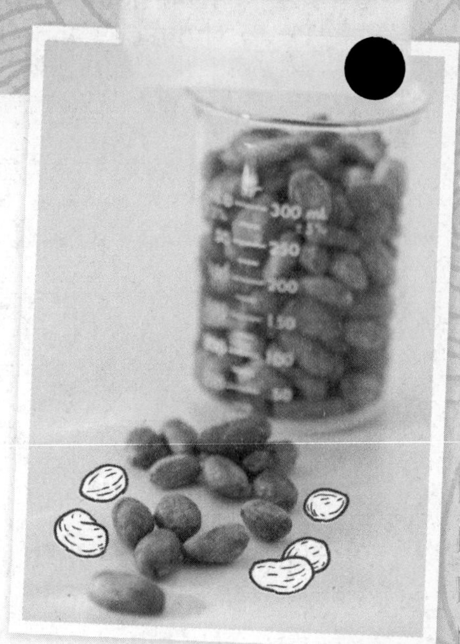

Safety

Materials

dried beans

beaker

Procedure

1. Read and complete a lab safety form.
2. Place 10 dried beans in a 100-mL beaker.
3. At the sounding of each start signal, you will double the number of beans in the beaker. Make a table to record your data. The table should indicate the number of beans added and the total number of beans in the beaker after each addition.

4. Double the number of beans each time the start signal sounds. Take time to record your data between each signal.
5. Follow your teacher's instructions for proper cleanup.

Analyze and Conclude

6. Can you add any more beans to the beaker? Why or why not?

7. How many times did you have to double the beans to fill the beaker?

8. Assess the following statement: The growth of a population does not affect the availability and use of resources, such as space or land. Use evidence from the lab to support your argument.

Population Limits As the human population grows larger and larger, cities become crowded with people, using space and depleting resources. Earth has limited resources. It cannot support a population of any species in a given environment beyond its carrying capacity. **Carrying capacity** is the largest number of individuals of a given species that Earth's resources can support and maintain for a long period of time. Let's explore carrying capacity.

Population Simulation

Kiwi Island is an imaginary 30,000-acre island in the South Pacific. The population of Kiwi Island was stable at 230 people between 2000 and 2010. Many areas of the island were tropical forests uninhabited by people. As more and more people discovered this paradise, the population began to double each year after 2010.

1. Plot the population growth of Kiwi Island from 2000 to 2017.

2. **MATH Connection** Determine the number of acres available per person for each year that is plotted.

3. **MATH Connection** Describe the relationship between land availability in 2000 and 2017.

4. Predict what happens to resources, such as land, if the population is below its carrying capacity. Explain your reasoning.

5. Predict what happens to resources if the population is reaching or above its carrying capacity. How might this impact the population growth of the island?

Approaching Carrying Capacity Each person uses space and resources. Population size depends on the amount of available resources and how members of the population use them. When population density is low, resources are abundant, and population increases. If resources become scarce or if the environment is damaged, members of the population can suffer and population size can decrease. If the human population continues to grow beyond Earth's carrying capacity, eventually Earth will not have enough resources to support humans. Let's investigate the relationship between resource use and population growth.

INVESTIGATION

Resource Consumption

The graph on the right shows the estimated amount of land required to support a person through his or her lifetime. This includes the land for the production of food, forest products and housing, and the additional forest land required to absorb the carbon dioxide produced by the burning of fossil fuels. A hectare of land is equivalent to approximately 2.5 acres or 10,000 square meters. Take a moment to compare and contrast the estimated hectares of land required to support individuals in each of the listed countries.

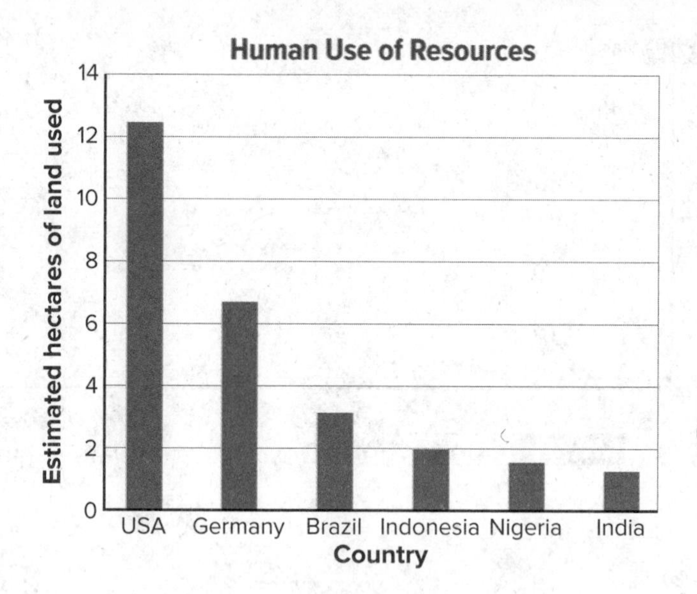

1. In which country is the per-capita consumption the greatest? In which country is the per-capita consumption the least? Why do you think human use of resources varies among countries?

2. **MATH Connection** Now, calculate the population growth rate (PGR) for each of these countries. Record this information in the table below. Use the following equation.

$$\frac{\text{birth rate} - \text{death rate} + \text{net migration rate}}{10} = \text{PGR (\%)}$$

Population Growth Rates

Country	Population 2015 (millions)	Birth rate	Death Rate	Net Migration Rate	PGR (%)
*United States	321.2	13	8	3	
*Germany	81.1	8	11	5	
Brazil	204.5	15	6	0	
Indonesia	255.7	21	6	–1	
Nigeria	181.8	39	14	0	
India	1,314.1	21	7	–1	

*Industrially developed country

Source: 2015 Population Reference Bureau

3. How do population growth rates compare in industrially developed countries and developing countries?

4. Compare this data with the Human Use of Resources graph on the previous page. Then evaluate the following statement: Countries with higher population growth rates have a greater impact on resource use than countries with lower population growth rates. Support or refute your argument with evidence.

5. Some developing countries, such as India, are becoming more industrialized and are also increasing in population. Predict the effect that this might have on the carrying capacity of Earth.

Growing Populations and Resource Use Another important factor in keeping the human population at or below the carrying capacity is the amount of resources from the biosphere that are used by each person. Currently, individuals in industrially developed countries use far more resources than those individuals in developing countries. Countries such as India are becoming more industrialized, and they have a relatively high growth rate. These countries are adding more people and are increasing their use of resources.

At some point, the land needed to sustain each person on Earth might exceed the amount of land that is available. Resources become harder to extract and find. However, a decline in these resources often leads to development of new technologies and alternative resources, which in turn expands Earth's carrying capacity. This allows for the population to increase again. But there is a cost to increasing populations. In the following lesson, we will examine how increases in human populations and per-capita consumption of natural resources impact Earth's systems.

THREE-DIMENSIONAL THINKING

Write a paragraph **explaining** why it is important for humans to maintain a population at or below the carrying capacity. Write your response in your Science Notebook.

COLLECT EVIDENCE

What impact does population growth and per-capita consumption have on Earth's resources? Record your evidence (B) in the chart at the beginning of the lesson.

STEM Careers

A Day in the Life of a Demographer

Have you ever heard the phrase *population explosion?* Population explosion describes the sudden rise in human population that has happened in recent history. The population has increased at a fairly steady rate for most of human history. In the 1800s, the population began to rise sharply. What kind of scientist studies this information?

A demographer is a scientist who studies populations, particularly human beings. Statistical data such as birth rates, death rates, migration rates, and occurrences of disease help these scientists illustrate the changing size, structure, and distribution of human populations.

A demographer also makes observations about the causes and effects of population changes based on collected data. For example, if an area is experiencing an increase in birth rates or immigration, they will analyze trends and predict future needs for the community.

It's Your Turn

Research Imagine you are a demographer for your city. Research how the population of your city has changed over the past 50 years. Create a short slideshow to predict how the needs of your community have changed based on the data you collect.

Summarize It!

1. **Organize** Complete the Venn diagram below. Below each term write its definition. Where circles overlap, explain the relationship among the terms.

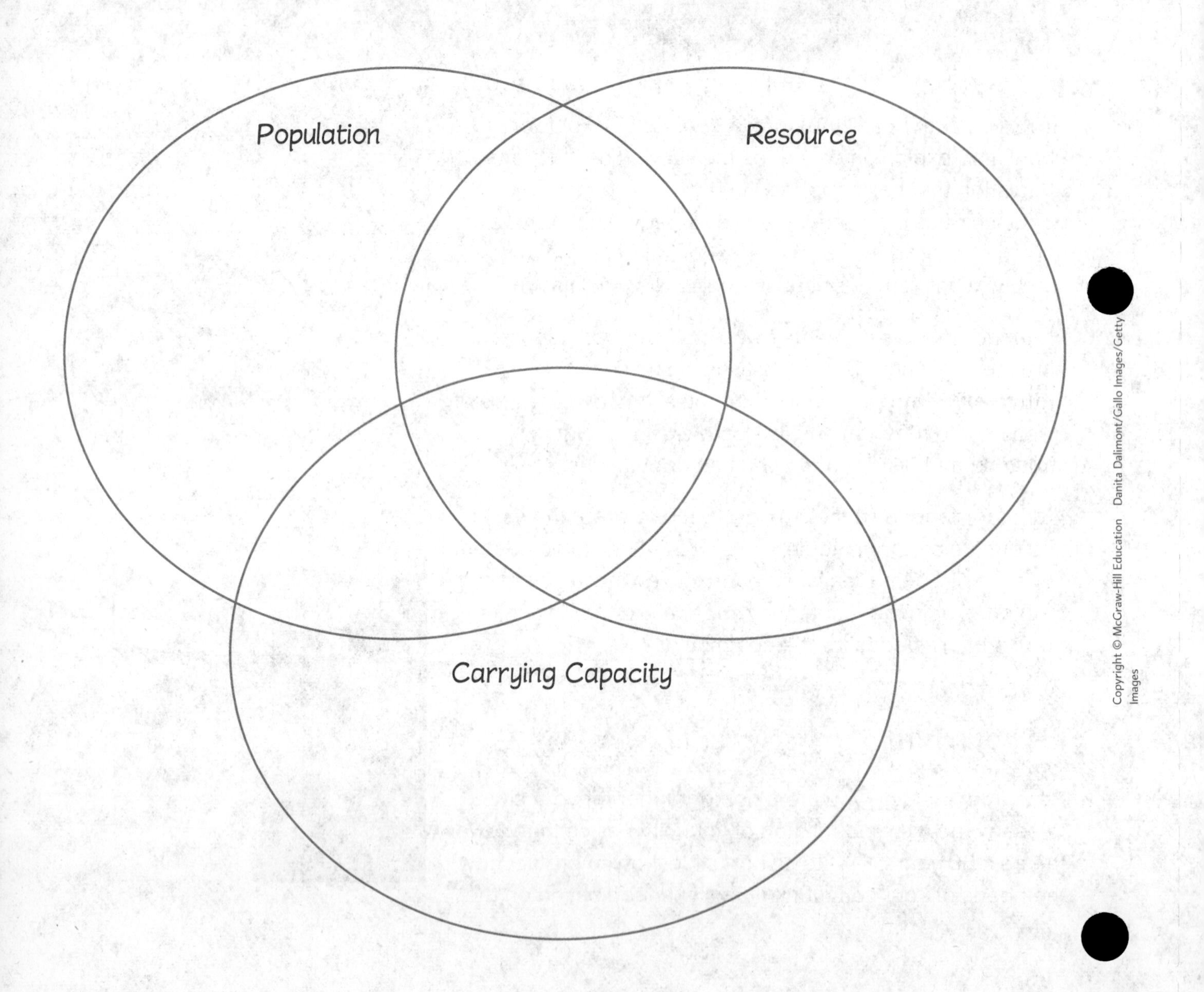

Three-Dimensional Thinking

Demographers often use age structure diagrams to study human populations in different countries. The shapes of these diagrams provide clues about how quickly the populations are growing. Compare the age distribution of males in the United States and Mexico.

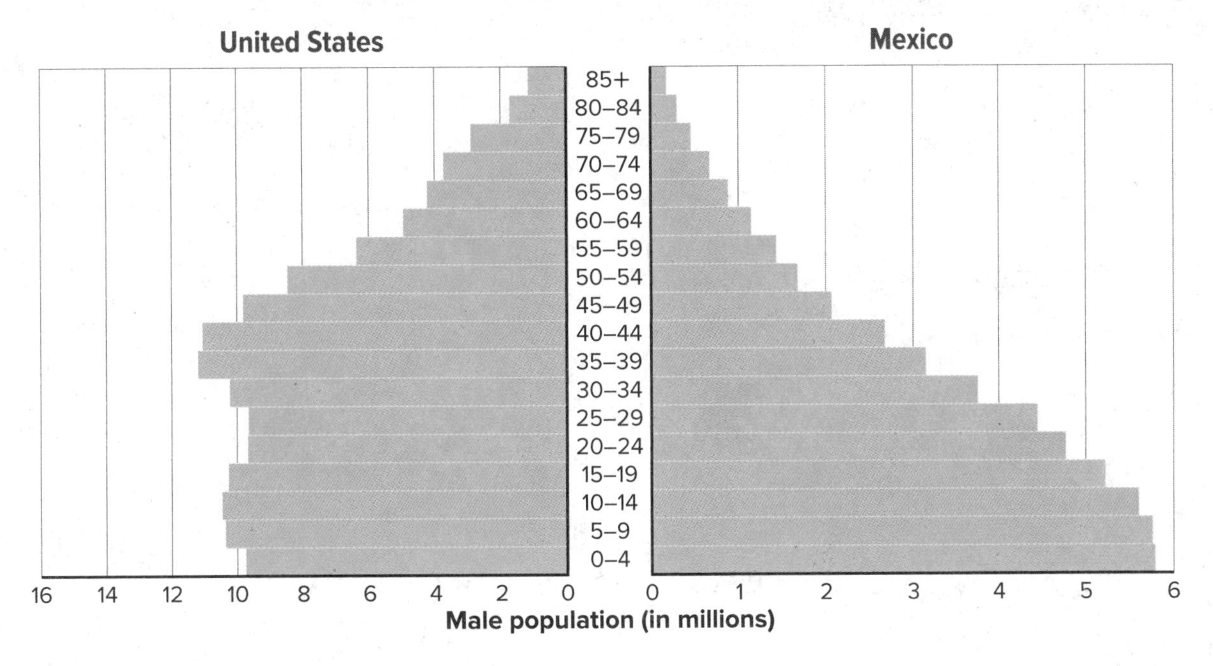

2. What can be concluded from the data displayed on the age structure diagrams above?

A The male population of Mexico is growing at a slower rate than the United States because fewer children are being born in Mexico.

B The male population of Mexico is growing at a faster rate than the United States because the number of children in Mexico exceeds the other age groups.

C The male population of Mexico is growing at a slower rate than the United States because there are fewer elderly males.

D The male population of Mexico is growing at a faster rate than the United States because there is a greater number of elderly males.

Real-World Connection

3. **Argue** A friend makes the following claim: There are enough resources on Earth to support increases in per-capita consumption rates if human population growth rates stay the same. Do you agree? Use evidence and reasoning to support or refute your friend's claim.

Still have questions?
Go online to check your understanding about human population growth.

Do you still agree with the student you chose at the beginning of the lesson? Return to the Science Probe at the beginning of the lesson. Explain why you agree or disagree with that student now.

EXPLAIN THE PHENOMENON

Revisit your claim about the relationship between resource availability and human population growth. Review the evidence you collected. Explain how your evidence supports your claim.

START PLANNING

STEM Module Project Science Challenge

Now that you have learned about human population growth and resource availability, go to your Module Project to begin preparing your argument. Keep in mind that you want to argue how the per-capita consumption of natural resources are impacted by human population growth.

LESSON 2 LAUNCH

Natural Resources

Four friends argued about natural resources and their impact on the environment. This is what they said:

Kate: It is better to use natural resources because they don't harm our environment like human-made resources.

Clint: It is better to use human-made resources because they don't harm our environment like natural resources.

Abby: It doesn't matter—both natural and human-made resources can harm the environment.

Troy: Neither human-made nor natural resources are harmful. They are both good for the environment.

Which friend do you agree with the most? Explain why you agree.

You will revisit your response to the Science Probe at the end of the lesson.

LESSON 2

People and the Environment

ENCOUNTER THE PHENOMENON | How does cutting down trees impact Earth's systems?

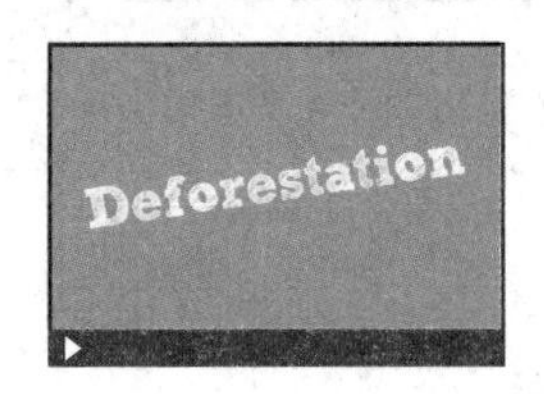

GO ONLINE
Watch the video *Deforestation* to see this phenomenon in action.

You have probably seen a tree lying on the forest floor that fell from natural causes. Trees are also cut down for human purposes. It may seem obvious that cutting down trees affects the forests and the other organisms that live there, but did you know each of Earth's systems is impacted when forests are removed? John Muir, an early naturalist for whom Muir Woods is named, once stated: "Tug on anything at all and you'll find it connected to everything else in the universe." With your partner, brainstorm how activities like cutting down forests impacts each of Earth's systems, and how we might mitigate these effects.

EXPLAIN THE PHENOMENON

Are you starting to get some ideas about how removing forests impacts Earth? Use your observations about the phenomenon to make a claim about how the consumption of resources impacts Earth's systems and how we might mitigate these impacts.

CLAIM

The consumption of natural resources, like trees, impacts... and can be mitigated by...

COLLECT EVIDENCE as you work through the lesson.

Then return to these pages to record your evidence.

EVIDENCE

A. What evidence have you discovered to explain why humans remove entire forests?

B. What evidence have you discovered to explain how the removal of trees impacts Earth's systems?

MORE EVIDENCE

C. What evidence have you discovered to explain how we can mitigate these impacts?

When you are finished with the lesson, review your evidence. If necessary, based on the evidence, revise your claim.

REVISED CLAIM

The consumption of natural resources, like trees, impacts... and can be mitigated by...

Finally, explain your reasoning for how and why your evidence supports your claim.

REASONING

The evidence I collected supports my claim because...

Why cut down trees?

Many experts are concerned about the loss of forest cover worldwide. The removal of large areas of forests for human purposes is called **deforestation.** Why do humans cut down trees?

INVESTIGATION

Falling Forests

1. **MATH Connection** In 1630 at the beginning of European settlement, about 425,000,000 hectares of the United States were forested. By 2017, the area of forested land had declined to 300,000,000 hectares. What is the percent decrease of forested land between 1630 and 2017?

2. **MATH Connection** The population of the United States in 1630 was around 4,500. In 2017, the population was around 326,500,000. What percent increase does this represent?

3. What cause-and-effect relationship can you identify between the removal of forests and human population growth?

Want more information?
Go online to read more about how human populations impact Earth's systems.

FOLDABLES®
Go to the Foldables® library to make a Foldable® that will help you take notes while reading this lesson.

Urbanization As you have discovered, increases in the size of the human population cause increases in the consumption of natural resources. Trees are cut down for fuel, paper products, and wood products. People also clear forests for urban development. **Urbanization**—the development of towns and cities on previously natural areas—is the most human-dominated and fastest-growing type of land use worldwide.

Agriculture Exploding population growth on our planet and the amount of land available to produce food for that population are not keeping pace with each other. Forest clearing for agriculture helps answer the question of how the world's growing human population can be fed. Brazil's Cerrado represents one of the world's last opportunities to open a large area of new, highly productive cropland. The Cerrado, 2 million hectares of grassland and tropical forest, is the site of the world's fastest-growing soybean production.

Rapid growth of soy production in Brazil has both positive and negative aspects. On one hand, more food is now available to feed the world. On the other hand, the rapid expansion of agriculture worldwide—at the expense of forests—has environmental effects on Earth's systems.

COLLECT EVIDENCE

Why do people remove entire forests? Record your evidence (A) in the chart at the beginning of the lesson.

How does deforestation impact Earth's systems?

Nearly half of the world's tropical deforestation occurs in the "arc of deforestation" between the Cerrado and the Amazon rain forest. Let's take a closer look.

INVESTIGATION

Arc of Deforestation

Study the map below. Then answer the questions that follow.

1. The continent's highest rate of forest clearing is in the "arc of deforestation." Why do you think these forests are cleared? Predict how this might affect the geosphere.

2. **LIFE SCIENCE Connection** Biologically, the Cerrado is the richest savanna in the world, with at least 130,000 plant and animal species. The Amazon rain forest is estimated to host some 10 percent of the planet's species. Predict the effects that agricultural expansion might have on the plant and animal species that live there.

3. Where on the map does deforestation strongly impact the hydrosphere? Explain.

4. Finally, predict how a reduction in trees might impact the atmosphere. Record your thoughts below.

Changes in Earth's Systems The consumption of natural resources, like trees and land, has environmental effects of global importance. Forest clearing for cropland and pasture diminishes habitat. Habitat loss is often the greatest extinction threat for the biosphere.

Clearing forests can also affect the geosphere and hydrosphere. Plant roots hold soil in place. Without these natural anchors, soil erodes away. Drought can also follow deforestation. Cleared land dries rapidly and stores little moisture.

And finally, deforestation also impacts the atmosphere. Trees remove carbon dioxide from the atmosphere during photosynthesis. Rates of photosynthesis decrease when large areas of trees are cut down, and more carbon dioxide remains in the atmosphere.

SEP DCI CCC

THREE-DIMENSIONAL THINKING

Outline an **argument** for how deforestation impacts Earth's **systems.** What are the short-term consequences of clearing land? Using your knowledge of the **cause-and-effect** relationships associated with deforestation, predict the long-term consequences of clearing entire forests.

COLLECT EVIDENCE

How does deforestation impact Earth's systems? Record your evidence (B) in the chart at the beginning of the lesson.

GREEN SCIENCE

Deforestation and Carbon Dioxide in the Atmosphere

How does carbon dioxide affect climate?

What do you think when you hear the words *greenhouse gases?* Many people picture pollution from automobiles or factory smokestacks from the consumption of energy resources. It might be surprising to learn that cutting down forests affects the amount of one of the greenhouse gases in the atmosphere—carbon dioxide.

Trees, like most plants, carry out photosynthesis and make their own food. Carbon dioxide from the atmosphere is one of the raw materials of photosynthesis. When deforestation occurs, there are fewer trees to remove carbon dioxide from the atmosphere. As a result, the level of carbon dioxide in the atmosphere increases.

Trees affect the amount of atmospheric carbon dioxide in other ways. Large amounts of carbon are stored in the molecules that make up trees. When trees are burned or left to rot, much of this stored carbon is released as carbon dioxide. This increases the amount of carbon dioxide in the atmosphere.

Carbon dioxide in the atmosphere has an impact on climate. Greenhouse gases, such as carbon dioxide, increase the amount of the Sun's energy that is absorbed by the atmosphere. They also reduce the ability of thermal energy to escape back into space. So, when levels of carbon dioxide in the atmosphere increase, more thermal energy is trapped in Earth's atmosphere. This can lead to climate change.

▲ These cattle are grazing on land that was once part of a forest in Brazil.

▲ In a process called slash-and-burn, forest trees are cut down and burned to clear land for agriculture.

It's Your Turn

WRITING Connection What are the short-term and long-term consequences of climate change? Draw evidence from research to support your analysis. Create a video blog in which you cite specific textual evidence to reinforce your argument.

What can be done?

Can Earth sustain our current lifestyles? Will there be adequate natural resources for future generations? These questions are among the most important in environmental science today. We depend on nature for food, water, energy, oxygen, and other systems that support life. For resource use to be sustainable, we cannot consume resources faster than nature can replenish them. Degrading ecological systems ultimately threatens everyone's well-being. Therefore, scientists, governments, and concerned citizens around the world are working to identify environmental problems, educate the public about them, and help find solutions.

ENGINEERING INVESTIGATION

Engineering Solutions

Technologies, such as the conservation drone pictured here, can help monitor and minimize human impact on the environment. Research two technologies used to reduce the effects of human populations on Earth's systems. Describe how these engineered solutions work in the space provided.

Call to Action Threats to supplies of food, energy, water, and nature certainly require technologies and action on national and international levels. However, in each of these cases, ethically responsible action can also begin with the individual.

Personal Choices The concept of an ecological footprint has been developed to help individuals measure their environmental impact on Earth. Your **ecological footprint** is defined as the area of Earth's productive land and water required to supply the resources that an individual demands as well as to absorb the wastes that the individual produces. You can calculate your own footprint by answering a simple questionnaire about consumption patterns, such as electricity use, shopping, and driving habits.

INVESTIGATION

Step Toward a Better Future

1. Calculate your ecological footprint score using the quiz provided by your teacher.

2. How many planet Earths does it take to support your lifestyle?

3. In the space below, brainstorm actions you could take to minimize your ecological footprint.

4. **HISTORY Connection** Think about the words of anthropologist Margaret Mead, who stated, "Never doubt that a small group of thoughtful, committed citizens can change the world. Indeed, it's the only thing that ever has." What do you think Mead was trying to draw attention to?

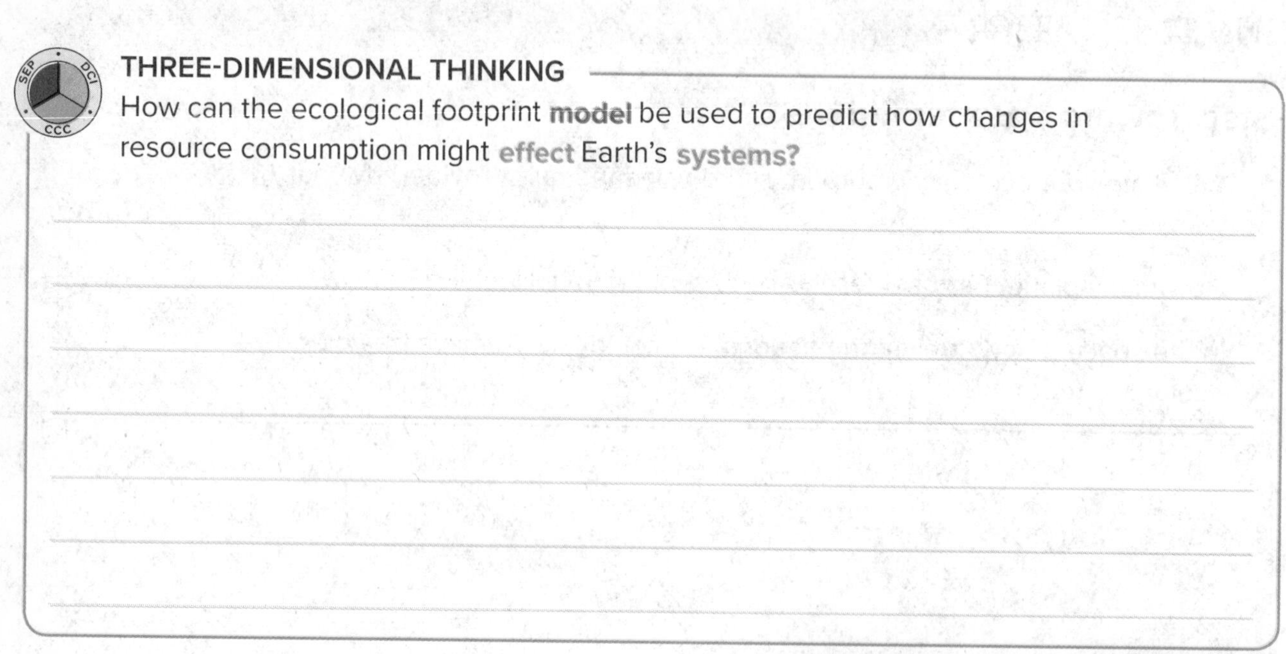

THREE-DIMENSIONAL THINKING

How can the ecological footprint **model** be used to predict how changes in resource consumption might **effect** Earth's **systems?**

Sustainable Resource Use Technological change sometimes can reduce our ecological footprint. For example, world food production has increased about fourfold since 1950, mainly through advances in irrigation, fertilizer use, and higher-yielding crop varieties, rather than through increased croplands. Similarly, switching to renewable energy sources such as wind and solar power can greatly reduce our ecological footprint. In Germany, which has invested heavily in wind, solar, small-scale hydropower, and public transportation, the ecological footprint is only 4.6 global hectares (gha) per person. In comparison, the average resident of the United States lives at a consumption level that requires 7.2 gha of biologically productive land.

COLLECT EVIDENCE

What types of technologies and activities help mitigate the effects of changes in Earth's systems? Record your evidence (C) in the chart at the beginning of the lesson.

Read a Scientific Text

You have learned that as the population increases, so does our impact on the planet. However, while science can describe the consequences of actions, science cannot necessarily prescribe the decisions that society takes.

CLOSE READING

Inspect

Read the passage *Should Environmental Scientists Be Advocates for Environmental Policy?*

Find Evidence

Reread the second paragraph. Highlight the arguments for and against having environmental scientists as advocates for environmental policy.

Make Connections

Communicate What are your thoughts? With your partner, discuss whether you think environmental scientists should be policy advocates. Does involvement in policy issues compromise the role of a scientist or enhance it?

Should Environmental Scientists Be Advocates for Environmental Policy?

Should environmental scientists be advocates for environmental policy? This question has been at the heart of policy debates since the environment became a major policy agenda item. A growing number of environmental scientists are now answering "yes." They argue that scientists, by virtue of being citizens first and scientists second, have a responsibility to advocate to the best of their abilities and in a justified and transparent manner.

Much of what has been written about advocacy looks at its appropriateness without adequately assessing its nature. For example, most of the arguments, whether they are for or against advocacy, often look at only one side of the debate. The question of scientific credibility is often raised. Those who oppose scientists acting as advocates often say that advocacy undermines a scientist's credibility. The counterargument could say that as long as a scientist's work is transparently honest, the scientific community is obligated to support it.

There is one area about which many scientists do agree: that as citizens first and scientists second, scientists have a responsibility to use their scientific data and insights to guide policy decisions. A growing number of scientists are calling for more active participation of their profession in matters of policy. They argue that broad participation of scientists in policy issues will undoubtedly result in disagreement among good scientists and will complicate the policy-making process. It is further argued, however, that the ultimate goal should not be simplicity in the process but rather the betterment of society. Many scientists state that the risk of not participating in policy debates on environmental issues is greater than the risk of participating.

LESSON 2

Review

Summarize It!

1. **Sketch** Create a concept sketch that describes your understanding of how changes in human populations have a causal role in changing Earth's systems. To construct a concept sketch, being by listing the relationships you want to describe. Then, draw your sketch and write complete sentences describing the sketch. Be creative!

Three-Dimensional Thinking

Biologists report that habitat destruction, overexploitation, and pollution are eliminating species at a rate comparable to the great extinction that marked the end of the age of dinosaurs. The UN Environment Programme reports that, over the past century, more than 800 species have disappeared and at least 10,000 species are now considered threatened. This includes about half of all primates and around 10 percent of all plant species. At least half of the forests existing before the introduction of agriculture have been cleared, and much of the diverse "old growth" on which many species depend for habitat is rapidly being cut and replaced by ecologically impoverished forest plantations.

2. Which statement is a logical conclusion that can be drawn from information in the text?

A Clean technology has helped eliminate pollution and protect endangered species.

B Natural resources are limited, and the biosphere has limited capacity to provide our food and water.

C Consuming forest resources interferes with the survival of other species.

D "Old-growth" forests supply food, energy, and other resources humans depend on.

---

3. **MATH Connection** Only a small percentage of Americans owned cars before the 1940s. By 2017, there were nearly 250 million vehicles for 323 million people, greatly increasing the need for roadways. In 1960, the United States had about 16,000 km of interstate highways. Today, the interstate highway system includes 77,000 km of paved roadways. What percent increase does this represent?

A 381 percent

B 792 percent

C 38 percent

D 79 percent

Real-World Connection

4. **Evaluate** the impact on land, water, and/or the atmosphere of an activity you perform daily that requires the consumption of natural resources. How could you minimize these effects?

Still have questions?
Go online to check your understanding about how human populations impact Earth's systems.

REVISIT

Do you still agree with the friend you chose at the beginning of the lesson? Return to the Science Probe at the beginning of the lesson. Explain why you agree or disagree with that friend now.

Revisit your claim about how natural resource consumption causes changes in Earth's systems. Review the evidence you collected. Explain how your evidence supports your claim.

PLAN AND PRESENT

STEM Module Project
Science Challenge

Now that you've learned about how humans impact the environment, go back to your Module Project to continue planning your argument to present in a panel discussion with your team. Your goal is to explain how increases in the human population and consumption of natural resources impact Earth's systems.

PERFORMANCE EXPECTATION

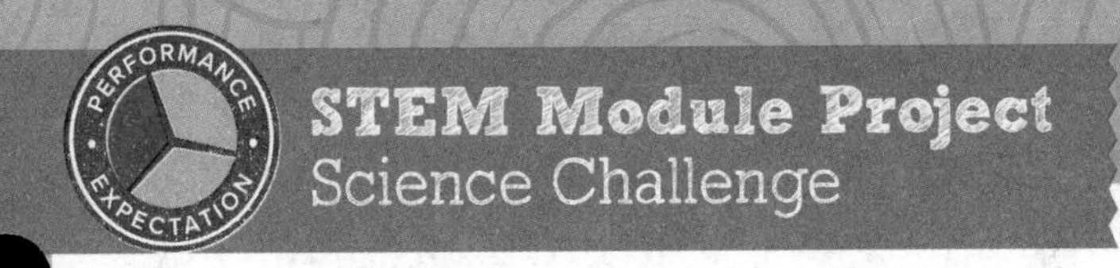

7.6 Billion and Counting

Scientists estimate that more than 7.6 billion people currently live on Earth. By 2050, the human population is projected to reach 9.8 billion. How do the human population and human activities impact Earth?

You are on a team of scientists who are trying to answer this question. You will gather evidence to support an argument about how increases in human population and resource use affect Earth's systems. You will present your argument in a panel discussion with your team.

Planning After Lesson 1

In the space below, list natural resources that you use every day. As a team, select a natural resource to investigate. Explain why you selected that particular natural resource.

STEM Module Project
Science Challenge

Planning After Lesson 1, continued

Next, select a particular region to research. In the space below, construct a graph that shows changes in the human population of the region over a given time.

Conduct research on the per-capita consumption of your resource by the human population in your chosen region. If possible, obtain information about past, current, and future rates of usage. Organize your findings in a chart or graph.

Compare your population data and your resource consumption data. Recognize and represent any patterns or relationships you see in your Science Notebook.

Planning After Lesson 2

How is your natural resource obtained or processed? How is it used?

How does the use of your resource change the appearance, composition, and/or structure of Earth's systems in your region? Include both direct and indirect impacts.

Research some solutions engineered to alter the impacts of the use of your resource on Earth systems. Describe the solutions in your Science Notebook.

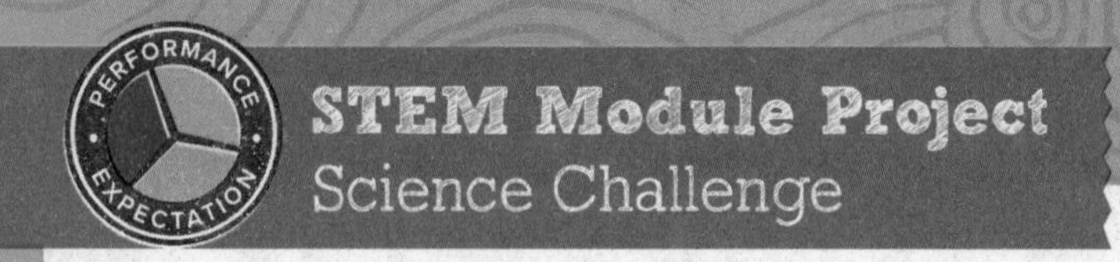

Construct Your Argument

Construct an argument supported by evidence and scientific reasoning for how increases in human population and per-capita consumption of natural resources impact Earth's systems. Include information about engineered solutions for these impacts.

Evaluate and Critique Your Evidence

After you write your scientific argument, take time to evaluate and critique the evidence you used to support your argument.

	Yes	No	Proposed Solution
Is your evidence necessary and sufficient for supporting your claim?			
Is your evidence sufficient to determine causal relationships between consumption of the resource and the impact on Earth's systems?			
Did you consider alternative interpretations of the evidence?			
Can you describe why the evidence supports your claim, as opposed to any alternative claims?			

STEM Module Project
Science Challenge

Present Your Argument

After you have written and evaluated your argument, answer the questions below. Then present your argument in a panel discussion with your team.

In the space below, summarize the relationship among the growth of human populations, resource use, and impacts on Earth's systems.

Review the solutions proposed to alter the effects of resource use on Earth's systems. Propose an improved solution to better reduce any negative effects on one or more Earth's systems.

Use what you've learned to evaluate this statement: Science can determine which decisions society makes about resource use.

Congratulations! You've completed the Science Challenge requirements!

Module Wrap-Up

REVISIT THE PHENOMENON

Using the concepts you've learned in the module, explain how patterns of light at night can reveal relationships among human population growth, resource use, and the resulting environmental impacts.

OPEN INQUIRY

What are one or two questions you still have about the phenomenon?

Choose the question you are most interested in. Plan and conduct an investigation to answer this question.

Multilingual Glossary

GO ONLINE to find multilingual glossaries for science.
The glossaries include the following languages.

Arabic	Korean	Tagalog
Bengali	Mandarin Chinese	Urdu
French	Portuguese	Vietnamese
Haitian Creole	Russian	
Hmong	Spanish	

Cómo usar el glosario en español:

1. Busca el término en inglés que desees encontrar.
2. El término en español, junto con la definición, se encuentran en la columna de la derecha.

Pronunciation Key

Use the following key to help you sound out words in the glossary.

a	back (BAK)	**EW**	food (FEWD)
ay	day (DAY)	**yoo**	pure (PYOOR)
ah	father (FAH thur)	**yew**	few (FYEW)
ow	flower (FLOW ur)	**uh**	comma (CAH muh)
ar	car (CAR)	**u (+ con)**	rub (RUB)
E	less (LES)	**sh**	shelf (SHELF)
ee	leaf (LEEF)	**ch**	nature (NAY chur)
ih	trip (TRIHP)	**g**	gift (GIHFT)
i (i + com + e)	idea (i DEE uh)	**J**	gem (JEM)
oh	go (GOH)	**ing**	sing (SING)
aw	soft (SAWFT)	**zh**	vision (VIH zhun)
or	orbit (OR buht)	**k**	cake (KAYK)
oy	coin (COYN)	**s**	seed, cent (SEED)
oo	foot (FOOT)	**z**	zone, raise (ZOHN)

English — A — Español

acid precipitation/climate change

precipitación ácida/cambio climático

acid precipitation: precipitation that has a lower pH than that of normal rainwater (pH 5.6).

precipitación ácida: precipitación que tiene un pH más bajo que el del agua de la lluvia normal (pH 5.6).

air pollution: the contamination of air by harmful substances including gases and smoke.

polución del aire: contaminación del aire por sustancias dañinas, como gases y humo.

aquifer: an area of permeable sediment or rock that holds significant amounts of water.

acuífero: área de sedimento permeable o roca que conserva cantidades significativas de agua.

C

carrying capacity: the largest number of individuals of one species that an ecosystem can support over time.

capacidad de carga: número mayor de individuos de una especie que un medioambiente puede mantener.

climate: the long-term average weather conditions that occur in a particular region.

clima: promedio a largo plazo de las condiciones del tiempo atmosférico de una región en particular.

climate change: any change in Earth's climate, including global trends in warming, cooling, precipitation, wind directions, and other related measures.

cambio climático: cambios de los patrones meteorológicos durante un periodo prolongado de tiempo.

conservation: the careful use of Earth's materials to reduce damage to the environment.

coral bleaching: the loss of color in corals that occurs when stressed corals expel the colorful algae that live in them.

conservación: uso cuidadoso de los recursos de la Tierra para reducir el daño al medio ambiente.

blanqueamiento de coral: pérdida de color en los corales que ocurre cuando los corales estresados expelen las algas de color que viven en ellos.

D

deforestation: the removal of large areas of forests for human purposes.

desertification: the development of desert-like conditions due to human activities and/or climate change.

deforestación: eliminación de grandes áreas de bosques con propósitos humanos.

desertificación: desarrollo de condiciones parecidas a las del desierto debido a actividades humanas y/o al cambio en el clima.

E

ecological footprint: the area of Earth's productive land and water required to supply the resources that an individual demands as well as to absorb the wastes that the individual produces.

huella ecológica: área productiva de la tierra y el agua de la Tierra requerida para proveer los recursos que un individuo demanda y también para absorber los desechos que el individuo produce.

G

global warming: an increase in the average temperature of Earth's surface.

greenhouse effect: the natural process that occurs when certain gases in the atmosphere absorb and reradiate thermal energy from the Sun.

greenhouse gas: a gas in the atmosphere that absorbs Earth's outgoing infrared radiation.

calentamiento global: incremento en la temperatura promedio de la superficie de la Tierra.

efecto invernadero: proceso natural que ocurre cuando ciertos gases en la atmósfera absorben y vuelven a irradiar la energía térmica del Sol.

gas de invernadero: gas en la atmósfera que absorbe la salida de radiación infrarroja de la Tierra.

N

natural resource: part of the environment that supplies material useful or necessary for the survival of living things.

recurso natural: parte del medio ambiente terrestre que proporcionan materiales útiles o necesarios para la supervivencia de los organismos vivos.

P

particulate matter: the mix of both solid and liquid particles in the air.

photochemical smog: air pollution that forms from the interaction between chemicals in the air and sunlight.

pollution: the contamination of the environment with substances that are harmful to life.

population: all the organisms of the same species that live in the same area at the same time.

partículas en suspensión: mezcla de partículas sólidas y líquidas en el aire.

smog fotoquímico: polución del aire que se forma de la interacción entre los químicos en el aire y la luz solar.

polución: contaminación del medioambiente con sustancias dañinas para la vida.

población: todos los organismos de la misma especie que viven en la misma área al mismo tiempo.

reclamation: a process in which mined land is restored with soil and replanted with vegetation.

recuperación: proceso por el cual las tierras explotadas se deben recubrir con suelo y se deben replantar con vegetación.

reforestation: process of planting trees to replace trees that have been cut or burned down.

reforestación: proceso de siembra de árboles para reemplazar los árboles que se han cortado o quemado.

U

urbanization: the development of towns and cities on previously natural areas.

urbanización: desarollo de los pueblos y las ciudades en zonas previamente naturales